Power Electronics and Power Systems

The Power Electronics and Power Systems Series encompasses power electronics, electric power restructuring, and holistic coverage of power systems. The Series comprises advanced textbooks, state-of-the-art titles, research monographs, professional books, and reference works related to the areas of electric power transmission and distribution, energy markets and regulation, electronic devices, electric machines and drives, computational techniques, and power converters and inverters. The Series features leading international scholars and researchers within authored books and edited compilations. All titles are peer reviewed prior to publication to ensure the highest quality content. To inquire about contributing to the Power Electronics and Power Systems Series, please contact Dr. Joe Chow, Administrative Dean of the College of Engineering and Professor of Electrical, Computer and Systems Engineering, Rensselaer Polytechnic Institute, Jonsson Engineering Center, Office 7012, 110 8th Street, Troy, NY USA, 518-276-6374, chowj@rpi.edu.

More information about this series at http://www.springer.com/series/6403

Prakash Ranganathan • Kendall E. Nygard

Distributed Linear Programming Models in a Smart Grid

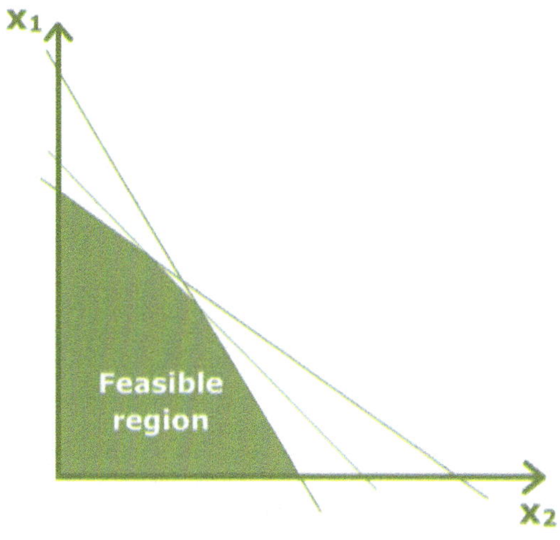

Springer

Prakash Ranganathan
Department of Electrical Engineering
University of North Dakota
Grand Forks, ND, USA

Kendall E. Nygard
Department of Computer Science
North Dakota State University
Fargo, ND, USA

ISSN 2196-3185 ISSN 2196-3193 (electronic)
Power Electronics and Power Systems
ISBN 978-3-319-84948-5 ISBN 978-3-319-52617-1 (eBook)
DOI 10.1007/978-3-319-52617-1

Printed on acid-free paper

This Springer imprint is published by Springer Nature
The registered company is Springer International Publishing AG
The registered company address is: Gewerbestrasse 11, 6330 Cham, Switzerland

"We dedicate this book to all smart-grid community researchers, and students who are working hard and making efforts to develop models for the next-generation power grid."

Prakash Ranganathan, Ph.D., &
Kendall E. Nygard, Ph.D.

Preface

A Smart Grid is a massively distributed and hierarchical electrical generation, consumption, and distribution system that is instrumented with sensors and utilizes many types of devices and controls. At a fundamental level, an electrical grid exists to provide the means of generating and routing power to meet consumption demands. The intent of a Smart Grid is to use technological advances in communication networks, sensing, and controls to manage the grid safely, reliably, and efficiently. Linear programming is a type of mathematical model that can explicitly produce solutions that optimize allocations of resources in the presence of constraints on their availability in time and place and on supporting infrastructure. Linear programming is also highly scalable to practical problems that are large and complex. Thus, this book concerns the use of Linear Programming models for resource-allocation problems that arise in Smart grid applications. The intended audience are members of the scientific community who work in energy-related fields, computer scientists, and engineering students. The book also serves as a reference book for anyone interested in the area of operations research for utility or other domains of interest.

In recent years, advances in computing and communication have resulted in large-scale, distributed environments. Environments that are capable of storing large volumes of data and, often, have multiple compute nodes. However, the inherent heterogeneity of the data components, the dynamic nature of distributed systems, the need for information synchronization and data fusion over a network, and the security and access-control issues make the problem of resource management and monitoring a tremendous challenge in the context of a Smart grid. Unfortunately, the concept of cloud computing and the deployment of distributed algorithms have been overlooked in the electric grid sector. In particular, centralized methods for managing resources and data may not be sufficient to monitor a complex electric grid. Most of the electric-grid management, including generation, transmission, and distribution, is, by and large, at a centralized control. In this book, we present a distributed algorithm for resource management which builds on the traditional simplex algorithm that is utilized to solve large-scale, linear optimization problems. The distributed algorithm is exact, meaning that its results are identical if it is run in a centralized setting.

More specifically, the authors discuss a distributed decision model, where a large-scale electric grid is decomposed into multiple submodels that can support the resource assignment, communication, computation, and control functions that are necessary to provide robustness and to prevent incidents such as cascading black-outs. The book's key contribution is to design, develop, and test a resource-allocation process through a decomposition principle in a smart grid. We have implemented and tested the Dantzig–Wolfe decomposition process with standard IEEE 14-bus and 30-bus systems. The book details how to formulate, implement, and test such an Linear Programming (LP)-based design in order to study the dynamic behavior and the impact of an electrical network while considering network's failure and repair rates. The Dantzig–Wolfe approach's computational benefits for finding an optimal solution and its applicability to IEEE bus systems are presented.

The last two chapters are dedicated to PMU placement problems and renewable energy allocation using the linear programming formulation. We hope you find this text useful to build research models using linear programming.

Grand Forks, ND Prakash Ranganathan
Fargo, ND Kendall E. Nygard

Acknowledgements

A number of people deserve thanks and acknowledgement for contributing to the completion of this work. First and foremost, I want to thank Professor Kendall E. Nygard, the second author of this book, for his guidance. Most of the book is based on my Ph.D. dissertation work which was advised by Kendall E. Nygard, Professor of Computer Science at North Dakota State University. Chapters 10 and 13 are based upon the work of my graduate students Ranganath Vallakati (M.S. student), Mitch Campion (M.S. student), and Arun Sukumaran Nair (Ph.D. student). Chapter 11 is based upon the work of Siby Plathottam (Ph.D. student), and this chapter was edited by Dr. Hossein Salehfar, Professor of Electrical Engineering at the University of North Dakota, and me. Damian Lampl, a Ph.D. student of Dr. Nygard at North Dakota State University, provided Chap. 12 and was edited by me. I sincerely thank my mentor, Dr. Kendall E. Nygard, and the students who have helped me in completing and adding meaning to this book. Finally, this book would not be possible without the support of my wife, Roopa, and daughter, Pragya, who have stood behind me throughout my career.

Thank you
Prakash Ranganathan, Ph.D.

Contents

List of Figures

List of Tables

List of Abbreviations

AMI	Automated Metering Infrastructure
AMPL	Algebraic Modeling and Programming Language
BB	Branch and Bound
CC	Control Center
DLP	Direct Linear Programming
DW	Dantzig–Wolfe Decomposition
DWLG	Dantzig–Wolfe Lagrangian Relaxation
GPS	Global Positioning Systems
IEEE	Institute for Electric and Electronic Engineers
LP	Linear Programming
MG	Micro Grid
MP	Master Problem
PDC	Phasor Data Concentrator
PMU	Phasor Measurement Unit
SP	Subproblem
ST	Subjected To
WAMS	Wide-Area Monitoring System

Chapter 1
Introduction

The worldwide electric-power industry is undergoing a transformation unlike anything that it has seen in over a century. The entire supply chain for electricity, including how the power is generated, transmitted, distributed, and consumed, is being overhauled with the goal of establishing a more sustainable energy future. Adopting new technologies and the associated market restructuring are a complex undertaking that requires knowledge about the many interacting variables and the conflicting cost functions for various market participants, such as power producers, system operators, load-serving entities, regulators, aggregators, service providers, and consumers. The smart grid is an information-enriched energy network, and it is going to require substantial information processing, storage, and data-mining resources. An entirely new software sector is being created to meet the challenges and to fill the many needs resulting from smart grid's arrival. Spending for the Smart grid is estimated to be $165 billion over the next 20 years, and a good portion of this cost will be for software and data services [RPT07, AW05]. The Smart grid is a complex, highly networked system that must operate in diverse, often-challenging environments that combine large and complex facilities with vast numbers of edge nodes, e.g., the smart meters that are grid's consumer-fronting boundary. The smart meters require sophisticated software in order to operate efficiently. Upgrading utility information and control infrastructure is critical to maintain the electric distribution system's reliability at a time of rising costs.

To meet the enormous challenge of creating a sustainable energy infrastructure for the future, which is driven by a smart grid, researchers and practitioners need to quantitatively investigate the complex interactions between different components of the electricity grid and to evaluate the impact of new ideas and technologies, considering the interdependencies among markets, power flows, and information and communication networks [Ami05]. To assist with a quantitative understanding of the grid, there is an unprecedented need for (near) real-time visibility about the state of the grid and its loads, with volumes of data being collected from smart meters and other sensing devices that are added to the grid. The book is a collection of design models which address these complex interactions through linear

P. Ranganathan, K.E. Nygard, *Distributed Linear Programming Models in a Smart Grid*, Power Electronics and Power Systems, DOI 10.1007/978-3-319-52617-1_1

Table 1.1 List of linear programming models

System models	Models
Model 1	An LP-based resource optimization that uses the Dantzig-Wolfe technique
Model 2	An optimal resource assignment uses a branch and bound model for a Smart grid
Model 3	A probabilistic energy-reallocation technique that uses linear programming in a Smart grid
Model 4	A placement problem of Synchro phasors using LP with redundancy criteria
Model 5	Unbiased optimum power flow model integrating wind sources
Model 6	A capacitated transshipment problem solver
Model 7	A distributed network decomposition into micro-grids using betweeness indices for economic dispatch

programming (LP) models. LP-based systems provide a paradigm for conceptualizing, designing, and implementing software systems with simplicity and robustness. The proposed, master LP model can act autonomously and can communicate with other LP structures across open and distributed environments.

The proposed approaches are comprised of multiple design models as shown in Table 1.1. The models are explained in the following chapters.

Model 3's contribution is a probability-based LP formulation for a directed network under uncertainty conditions with supply and demand units. Here, our contribution is modeling and expanding the basic integer linear-program formulation of a bi-partite graph to a network grid structure with known uncertainty. This approach is described in Chap. 3. Model 2's contribution involves developing and implementing a branch-and-bound technique that allocates the distributed energy resources (DERs) to a set of demand units. Here, we discuss how distributed energy resources can maximize their preferences when subjected to various equality and inequality constraints. Model 3's contribution is the formulation, implementation, and testing of a decomposition procedure for an electric-grid resource-allocation problem. Models 4 and 5 discuss a placement problem of PMUs and integrating wind sources to the power grid. Models 6 and 7 detail a capacitated transshipment solver and economic dispatch problem using LP methods.

In the book, we treat an agent as a piece of software code that can run LP functions continuously and autonomously in an environment where other processes take place and where other agents exist. The sense of "autonomy" means that the agent activities do not require constant human guidance or intervention. We envision this architecture as a distributed system consisting of a collection of autonomous micro-grids, which can make decisions themselves, connected through an electrical network and distribution middleware, which enables the Independent System Operator (ISO) to coordinate the activities and to share the smart-grid system's resources so that consumers perceive the system as a single, integrated computing facility.

Smart-grid technology promises to revolutionize the way electricity is produced, delivered, and utilized. A fundamental problem when building open-distributed systems is to design mechanisms that compute optimal system-wide solutions effectively despite the self-interest of individual micro-grids. In particular, using renewable energy sources is expected to result in a massively distributed power-generation and distribution system that is composed of a large number of generating stations which operate on disparate renewable technologies. The optimal allocation of existing energy resources becomes a challenge due to the massively distributed nature of the generation facilities and consumption sites, and due to the uncertainty caused by inherent random fluctuations with generation. How should resources be effectively and computationally allocated to such a highly distributed system? Therefore, we target a distributed resource-allocation problem to satisfy both the local and global objectives in order to reach an optimum solution for a smart-grid application by studying the current IEEE electric-grid bus systems. We propose an iterative, distributed algorithm for the resource optimum problem as a solution. The algorithm is scalable for deployment in large electricity networks because it requires fewer computations than modeling via a centralized, direct LP implementation.

1.1 Objectives of the Book

The main objective of this book can be summarized in a single sentence: to develop an LP model for the resource-allocation problem in a Smart grid. The book focuses on a distributed linear-programming technique for an electric utility's resource-allocation problem. The computational effectiveness of the Dantzig-Wolfe modeling, and the associated tasks and objectives are given in the following sections:

1.1.1 Objective #1. Formulate a Mathematical Model for the Smart-Grid Resource-Allocation Problem

Task 1: Study and review prior modeling approaches that are explained in the literature in order to ascertain the computational benefits for resource-allocation problems. To study how these approaches can be applied in a smart-grid application by reviewing various techniques, such as LP, fuzzy logic, and heuristic methods, the Literature Review is presented in Chap. 2. Compared to the other formulation types reported in the literature, the Dantzig-Wolfe (DW) LP formulation has a much simpler structure, and we argue that it can be modeled for large-scale systems such as the IEEE 30 bus system for a Smart grid. Solution algorithms for general LP problems exist in commercially available software; these solvers, however, are intended for generic problems and cannot detect and take advantage of special problem structures, limiting the size of the problem that can be solved. We address how Smart-grid resources can be formulated as a special case structure in order to apply the DW method.

Task 2: Formulate the Dantzig-Wolfe decomposition constraints for an IEEE 14- and IEEE 30-bus system. An LP formulation of inter-regional constraints of a large-scale grid by decomposition process is described in Chap. 5.

Task 3: Formulate the bi-directional flow network constraints.

Task 4: For the 14-bus system, decompose the entire grid into multiple regions (Regions 1, 2, and 3), and formulate their constraints.

Task 5: Develop a two-region decomposition formulation of the same problem, and compare the resulting optimal resource-allocated solutions with the three-region decomposition's formulation. This task provides information about whether all decompositions yield similar performance and computational time savings.

1.1.2 Objective #2. Design, Develop, and Implement a Distributed Solution Procedure for the Mathematical Model

Task 1: Develop any additional constraints and objectives for the proposed problem. Here, the objectives are twofold: (1) to reduce the overall system-failure rate and (2) to reduce the repair rate for an IEEE bus system.

Task 2: Study the suitable LP solver tools to implement such a scheme.

Task 3: Design a distributed solution procedure to compute dual values for the Dantzig-Wolfe procedure, and exchange dual values between local micro-grids and global objectives.

Task 4: Implement the LP approach directly, without decomposition, in a large-scale algebraic LP solver such as A Mathematical Programming Language (AMPL).

Task 5: Implement Task 4 with the decomposition applied.

1.1.3 Objective #3. Develop an Experimental Design for Testing the Procedure Referenced in Objective 2

Task 1: Set up the simulation environment in AMPL for the IEEE 14- and IEEE 30-bus systems.

Task 2: Develop experimental design parameters for power-flow constraints in the transmission lines. Restrict the flow in one direction at a time by using binary operators.

Task 3: Study the feasibility regions of the proposed mixed-integer problem.

Task 4: Develop contingency scenarios about how the method will react to, along with the feasibility of, the solution it provides.

Task 5: Choose the Computer Processing Unit (CPU) run time as one of the main performance parameters.

1.1.4 Objective #4. Conduct the Experimental Testing Referenced in Objective 3

Task 1: Test the experimental setup for various scenarios. Conduct sensitivity analysis by simulating line failures and observing the optimum.
Task 2: Compare the DW procedure with direct LP implementation, and analyze the resultant computational savings.
Task 3: Test the procedure's scalability as the number of resources and the system demand increase. For example, how does the procedure scale with a 30-bus system?

1.1.5 Objective # 5. Develop Decision Models Using Linear Classifier, and Placement of Synchro phasors Using LP

Task 6: This task discusses on how decision can be made using simple linear classifier, and a placement problem of a sensor known as Synchro phasor with zero injection constraints.

1.1.6 Objective # 6. Integrating Wind Source to Smart Grid Decision Using Linear Programming, and Modeling Capacitated Resources

Task 7: This task details how uncertainty in the wind resource can be reduced, when integrated to smart grid networks, and also discusses a Capacitated Transshipment Problem Solver. Chapter 12 in this task also outlines novel way to decompose micro-grids using between indices.

The tested results are detailed in Chap. 6, yielding significant results about the computational savings for such a decomposition procedure that is used by utility operators during contingency scenarios. Such a decomposition formulation is neither tested nor formulated using real IEEE bus-system data. The DW procedure's applicability is the first of its kind to consider bi-directional power-flow constraints. We strongly believe that this novel technique will enable local system operators to predict, apply, maintain, and balance the resource allocation effectively for their systems in a time-sensitive grid. We assert that this approach will generate broad interest in the utility market for analysis and adaptation. Moreover, the procedure is guaranteed to converge and does not require the revelation of local information from each micro-grid, and all algorithm actions can be realized by programmable smart devices on the smart grid.

Table 1.2 Objectives and the number of tasks

Completed objectives	Completed tasks	Chapters
Objective #1	Tasks 1, 2, 3, 4, and 5	Chaps. 1–7
Objective #2	Tasks 1, 2, 3, 4, and 5	
Objective #3	Tasks 1, 2, 3, 4, and 5	
Objective #4	Tasks 1, 2, and 3	
Objective #5	Task 6	Chaps. 8 and 9
Objective #6	Task 7	Chaps. 10–12

Table 1.2 shows the number of tasks that are completed for each objective. The book has 11 chapters that include this Introduction chapter. Chapter 2 details the Literature Review for linear programming and Smart-grid modeling. Two published papers are included as Chaps. 3 and 4, and they describe probabilistic resource-reallocation modeling as well as a branch-and-bound technique, respectively.

In summary, we have contributed to the three models using LP in the book. The first model contribution is a probability-based LP formulation for a directed network under uncertainty conditions with supply and demand units. Here, our contribution is modeling and expanding the basic integer linear-program formulation of a bi-partite graph to a network-grid structure with known uncertainty. This approach is described in the published paper given in Chap. 3. Our second contribution involves developing and implementing a branch-and-bound technique for allocating distributed energy resources (DERs) to a set of demand units. Here, we present results that show how distributed energy resources can maximize their preferences when subjected to various equality and inequality constraints. This model is completely developed, implemented, and tested by using a branch-and-bound method. This method is outlined in the published paper that is presented in Chap. 4.

Chapter 5 details a decomposition model using the Dantzig-Wolfe procedure as well as its formulation for the standard IEEE 14-bus and IEEE 30-bus systems. This procedure is the book's third major contribution. Chapter 6 describes the implementation and testing phase as well as the results of the Dantzig-Wolfe procedure. The conclusion and future work are provided at the end of this book. Chapters 6 and 7 discuss implementation, testing, and remarks on application of Dantzig-Wolfe decomposition for micro-grid decomposition.

Chapters 8 and 9 detail on a linear classifier, and placement of PMUs using zero injection constraints. Chapter 10 is a research work on optimal power flow for power systems with wind power integration using LP approach. Chapter 11 discusses a capacitated transshipment problem solver.

Chapter 2
Literature Review

The goal of this chapter is to provide prior work done with linear-programming approaches for the resource-allocation problem. Operations-research (OR) modeling often concerns finding the best quantitative solution for management problems [HL01, Mom01]. The OR methods include mathematical-optimization modeling as simulation, and using OR methods has grown significantly since their origination during World War II. Templeman [Tem91] describes quantitative OR methods for designing and controlling industrial and economical operations. Many private and government organizations have improved their operations by successfully using mathematical programming [Wad83, Aro02, Chv83, Dan63, SS85]. This book focuses on a resource-allocation problem and applies linear programming for the solution approach.

2.1 Linear Programming in Practice

LP problems are decision problems where the purpose is to compute values for a set of decision variables in order to optimize (maximize or minimize) a linear-objective function, subject to a set of linear constraints. A formal definition for the class of LP problems is given below; because this book is primarily about solving LP problems in practice, I, first, briefly consider the context in which such problems arise and the importance of being able to solve them.

A diverse range of real-world problems can be approximated and formulated as LP problems, and there is often great economic or other value attached to finding an optimal solution. The LP field was originally developed to plan military-logistic operations during the Second World War (The word "programming" in LP means "planning."), and since then, the range of applications has flourished. Examples include industrial diamond blending, hired-car fleet management, distribution warehousing and supply chain management, oil refining, and gas-pipeline flow.

P. Ranganathan, K.E. Nygard, *Distributed Linear Programming Models in a Smart Grid*, Power Electronics and Power Systems, DOI 10.1007/978-3-319-52617-1_2

(Many other applications are described in the literature [GPS00, BBG77, Bou01, Bou02, Wil93].)

The value of being able to identify an optimum solution, as opposed to a feasible solution or sometimes no solution, can run into the order of many millions of dollars. For instance, a difference of 1% in the objective value in the yearlong PowerGen problem represents an annual cost difference of $520 million [Pow98].

Dr. Warren Powell of Princeton University and others developed a model for the Commercial Transport Division of North American Van Lines. Under high levels of demand uncertainty, this model dispatches thousands of trucks from customer origins to customer destinations each week. Working closely with upper management, the project team developed a new type of network model for assigning drivers to loads. The model, LOADMAP, combined real-time information about drivers and loads with an elaborate forecast of future loads and truck activities in order to maximize profits and service. It gave management a new understanding about the economics of truckload operations; integrated load evaluation, pricing, marketing, and load solicitation with truck and load assignment; and increased profits by an estimated $2.5 million annually while providing a higher level of service [PSN88].

The growth of LP as a practical technique would not have been possible without simultaneous progress in the power and availability of computing. Today, software for LP optimization is highly sophisticated, with several commercial codes being actively developed and marketed. A symbiotic relationship exists between the capacity of the codes and the growth of applications, with solutions for larger problems being demanded in less time as codes improve. This book investigates a well-known, but not well-used, method which has the potential to solve large problems quickly by exploiting the structure.

LP optimization software uses two main method classes. The simplex method is a gradient-descent method that moves along the edge of the feasible region [Chv83, Dan63]. Interior-point methods (IPM) move through the interior of the feasible region [Wri97]. We do not dwell on these well-known methods but take the solution of an LP problem with either of these methods as granted, provided that practical considerations allow it. DW decomposition was developed in the late 1950s, a decade after the simplex method and many years before interior-point methods were applied to LPs [Dan63, Dan83]. The DW procedure immediately aroused widespread interest, and many attempts were made to implement it as a computational method. Practical experiences, however, were mixed, with some claims of success but no lasting achievements when measured by the methods used to solve practical problems. There has been no evaluation about and development of different options and strategies for computational implementations, whereas there have been continued research and development for over 50 years with the simplex method and for over 20 years with the Integer Programming Method (IPM). Perhaps the greatest challenge with the Dantzig-Wolfe decomposition has been that, when viewed simply as an alternative LP optimization method, successes have been rapidly overtaken by improvements in simplex and IPM implementations. The continued improvement with LP optimization technology could be used to enhance the implementation of the Dantzig-Wolfe decomposition.

In the book, we primarily focus on the technique referred to as Dantzig-Wolfe decomposition (DW), an optimization technique for solving large-scale, block-structured, linear-programming (LP) problems. Opportunities from many different fields, such as production planning, refinery optimization, and resource allocation, may be formulated as LP problems. Where there is some structure arising from repeated problem components, such as handling multiple periods, locations, or products, the problem may potentially be solved by using the Dantzig-Wolfe decomposition.

As preparation for our practical work, we investigate how suitable block structures can be identified in practical LP problems. We develop the decomposition algorithm from first principles, delineating the theoretical requirements and showing which aspects are open for experimentation in a practical implementation. We illustrate, geometrically, the transformation from the original problem to the Dantzig-Wolfe master problem, and we establish precisely how solutions obtained from the decomposition algorithm correspond to answers for the original problem. We critically review previous practical work.

Smart-grid control systems have used both centralized and decentralized (distributed) approaches [BCP08]. Centralized control systems have the best performance for small-scale power networks and delivering power in one direction (i.e., from substation to loads). Today, the evolution of some power-distribution routines, such as distributed power storage and distributed generators (DGs), requires deploying smart-control systems [NF12]. Most traditional power-control systems act preventively or reactively to events, whereas more recent control systems add active control options to their strategies [Wan01]. Control architectures for power grids have widely utilized central and hierarchal methods. Considering their higher efficiency and reliability, decentralized and fully distributed intelligent controllers are beginning to appear [DNS95].

Optimization techniques have been used for power systems and have been studied with many resource-allocation applications [Son99, Moo91, Sal04]. Power-distribution networks are usually designed radially with load-feed flows in one direction. This type of network experiences increased loss, decreased voltage amplitude, and voltage instability (when using a motorized maximum load) due to its radial design and, probably, its long length. One effective solution for improving the performance of distribution networks, from a technical point of view, is using distributed generation supplies. Generally speaking, the advantages of using a distributed generation pattern can be categorized into two technical and economic aspects [KHS05].

The technical advantages of a distributed generation include decreased line loss, improved voltage profile, decreased environmental pollution, increased energy efficiency, higher-quality power, improved system reliability, and security. On the other hand, the economic advantages of applying distributed generation patterns include various investments to improve facilities, decreased operational costs, optimized production, decreased costs to save energy, and increased security for critical loads. A distributed allocation technique using branch and bound is studied for allocating DERS in the context of a smart grid [RN10]. A probabilistic approach using linear programming is applied for a Smart-grid resource-allocation problem. Several other Smart-grid implementations for a self-healing grid using LP are studied [PFR09]. LP-based decision support for situational awareness is outlined. Comprehensive universal markup language (UML) representations of micro-grid architecture are developed. The preliminary results of the resource-allocation problem in a Smart grid using the Dantzig-Wolfe procedure are presented.

2.2 Development of a Distributed Linear-Programming Model

A massive power blackout that caused some five million people in Arizona, California, and Mexico to lose electricity was apparently triggered by one person in Arizona. Figure 2.1 shows a map of the electric outage areas. An Arizona Public Service worker "removed a piece of monitoring equipment" which set off a chain reaction across the region, according to the Associated Press [RN12]. The outage appeared to be related to a procedure that an Arizona Public Service (APS) employee was performing at the North Gila substation which is located northeast of Yuma. Operating and protection protocols typically would have isolated the resulting outage to the Yuma area. The reason that the isolation did not occur in this case was mostly blamed on a lack of automated programs and reliance on heavy manpower. Our approach addresses such outage events through the LP programs discussed in the next paragraph [DCN04, FES12, GPR09].

We restrict the attention to the general LP approach and the Dantzig-Wolfe decomposition technique in the context of a Smart grid. The following macro-grid architecture has a centralized agent called an independent system operator (ISO) which coordinates the micro-grid activities. An agent is a piece of software code that performs tasks autonomously in the event action is needed to restore the grid process, such as self-healing the grid during an outage, running resource allocation, or scheduling tasks [NF12]. Thus, every micro-grid has an objective function and constraints that are formulated as an LP problem. We treat each individual LP program as an individual agent that monitors these micro-grids and associated activities as shown in Fig. 2.2.

Similarly, any macro-operations, such as coordinating all micro-grids, are conducted by an independent system operator, and they are treated as a master LP program and constraints run by AMPL. The master LP program interacts with the sub-problems via the exchange of dual variables. Figure 2.3. shows the distributed linear programming architecture.

As an information infrastructure with monitoring, control, and protection functions in a smart-transmission grid, the wide-area measurement system (WAMS), based on a synchronized phasor measurement unit (PMU), gradually becomes an important guarantee for the security and stability of power systems. The WAMS can be used to conduct real-time monitoring and control for dynamic system states,

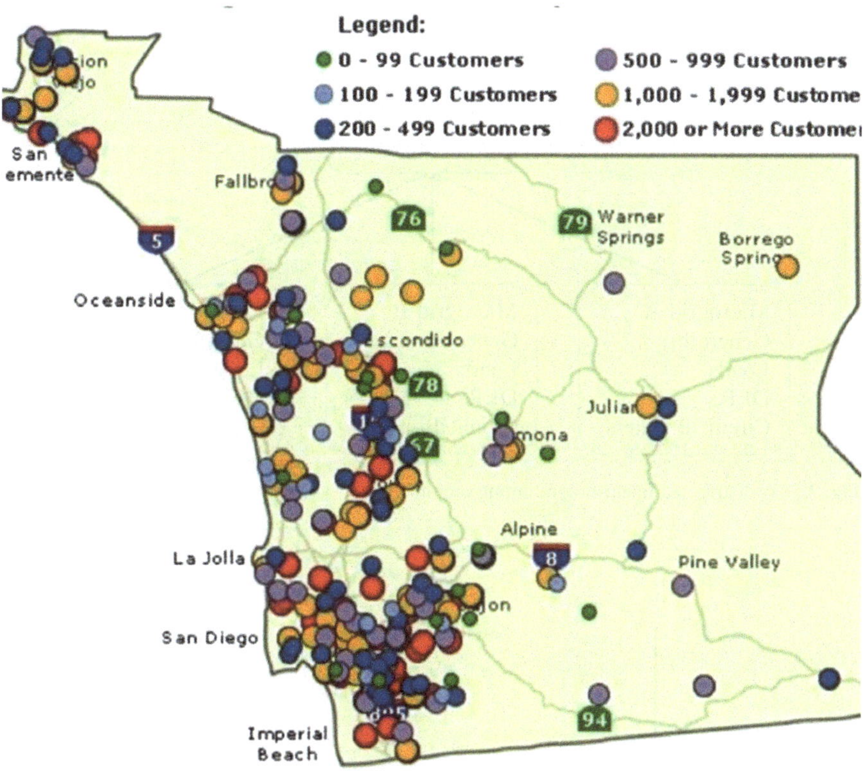

Fig. 2.1 Electric outages in the Sandiego Gas and Electric (SDG and E) Territory: September 8, 2011, 6:39 pm [RN12]

Fig. 2.2 LP as agents

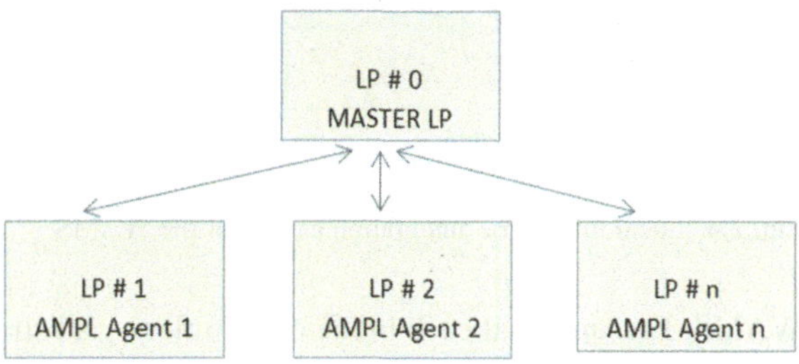

enhancing the system's security level because it utilizes the highly precise, synchronous clock in a global positioning system (GPS) to build unified time-space synchronization. The WAMS usually includes the PMUs, phasor data concentrator (PDC), control center (CC), and the high-speed data communication networks. Figure 2.4 shows a local micro-grid with PMU-PDC integration as part of the

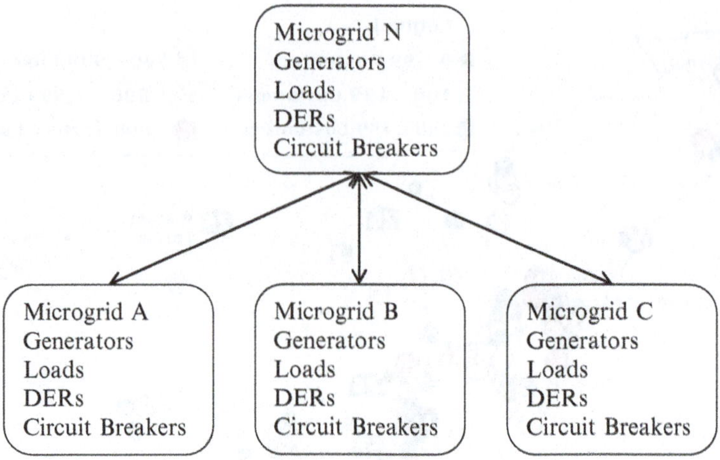

Fig. 2.3 A distributed linear-programming architecture.

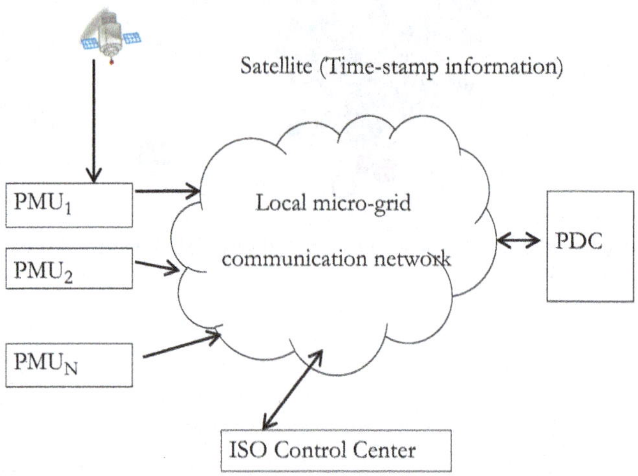

Fig. 2.4 Local micro-grid integration as part of the WAMS

WAMS. I assume that each block of the micro-grid structure shown in Fig. 2.4 has these units and integration in place.

Applying the Dantzig-Wolfe procedure would be significant if it is used with the WAMS or micro-grid architecture. In the book, we will show the computational significance of a small-scale grid, such as an IEEE bus network, to demonstrate its computational efficiency. Chapters 3 and 4 discuss a resource-allocation procedure where the preliminary results motivated us to continue the computational study of the Dantzig-Wolfe procedure.

Chapter 3
Energy Reallocation in a Smart Grid

When a malfunction occurs in a smart-grid electricity-provisioning system, it is vitally important to quickly diagnose the problem and to take corrective action. The self-healing problem refers to the need to take action in near real time in order to reallocate power to minimize the disruption. To address this need, we present a collection of integer linear programming (ILP) models that are designed to identify the optimal combinations of supply sources, the demand sites for generators to serve, and the pathways along which the reallocated power should flow. The models explicitly support multiple time periods and the uncertainty associated with alternative sources such as wind power. Model solutions are evaluated using a simulator configured with multiple, intelligent, distributed software agents.

3.1 Introduction

A Smart grid is a digital-age, electrical generation, and distribution system that is fully networked, instrumented, controlled, and automated. A Smart grid is a quintessential machine-to-machine system where the major components, such as generators, relays, transformers, power lines, and electrical meters, are networked and digitally addressable with methods such as Internet Protocol (IP). Many components are also equipped with sensors and processors that are capable of carrying out intelligent actions with little or no human intervention. Available

---

The material in this chapter was co-authored by Prakash Ranganathan and Kendall E. Nygard. Prakash Ranganathan had primary responsibility to develop the linear-programming formulation for a resource-allocation problem that includes flow-balance constraints and uncertainty information. Prakash Ranganathan was the primary developer of the modeling conclusions that are advanced here. Prakash Ranganathan also drafted and revised all versions of this chapter. Kendall E. Nygard served as a proofreader and checked the LP formulation conducted by Prakash Ranganathan. The others contributed to software interface.

P. Ranganathan, K.E. Nygard, *Distributed Linear Programming Models in a Smart Grid*, Power Electronics and Power Systems, DOI 10.1007/978-3-319-52617-1_3

power resources for a Smart grid include conventional types of generating plants and small-scale, renewable distributed energy resources (DERs).

A Smart grid provides great potential advantages for many stakeholders. At the user level, smart-meters at power-demand sites create possibilities for dynamically pricing electricity, making it possible for consumers to receive lower rates by shifting their usage from periods of high demand to times of low demand. Smart meters also assist utilities by reducing peak loads, allowing meters to take action to optimize resource allocation and to maximize efficiency. When disruptions occur, the grid's instrumentation immediately communicates exact information that pinpoints the location and type of problem, making maintenance and repair activities more responsive and efficient. At the transmission-grid level, PMUs placed at strategic locations provide detailed information about grid health, and can trigger messages that report problems or initiate control actions.

Cascading failures that have occurred in past years highlight the need to understand the complex phenomena that can occur in power networks and to develop emergency controls and restoration procedures. In addition to mechanical failures, overloading a line can create power-supply instabilities such as phase or voltage fluctuations. A truly intelligent grid is able to predict impending fault states and failures [ADH94, CLD02, AS08, DCN04].

Self-healing capabilities are highly desirable in a Smart grid. We define self-healing as the ability to detect the need for corrective actions in the grid and to autonomously carry out such actions. Once a fault state is detected, the grid should perform appropriate procedures, such as dynamically controlling the power flow to restore grid components from a fault state to normal operation. Examples of common failures that occur in the power grid are power outages, low power quality, overloads that could lead to cascading failures, and service disruptions.

In our work, we model the topology of the Smart grid as an abstract network of nodes representing supply sources, demand sites, and transshipment junctions that are interconnected by links representing transmission lines. Devices such as generators, relays, and transformers are associated with specific nodes. Our models are integer linear programs that provide a self-healing capability by identifying optimal alternatives for reallocating and rerouting power when disruptions and failures occur. Failures affect the ability of supply sources to meet energy demands at certain sites. Our primary modeling goal is to balance the flow of power across the system to ensure that no consumer site experiences an outage while also maximizing the system's overall efficiency, cost-effectiveness, and reliability. Our models account for multiple factors, such as availability, reliability, uncertainty, cost-effectiveness, and consumer preference. The basic modeling template is the Capacitated Transshipment Problem (CTP). An additional model structure incorporates uncertainty at supply sources and ensures that capacities (load limits) for the transmission lines and through devices are not exceeded. Uncertainty about the available supply at certain sources is modeled within the integer linear-programming framework using chance-constrained programming methods. The integer linear-programming models provide the basis for intelligent decision-making in the grid as model pertains to resource allocation. An agent-oriented simulation of the Smart-grid operation is available to test and evaluate alternative resource-allocation solutions.

This chapter is organized in six sections following the Introduction. Section 3.2 provides the Problem Statement and necessary background. Section 3.3 provides a brief review of Smart grid modeling and ILP. In Sect. 3.4, we present a collection of ILP models that capture various aspects of the self-healing problem, including an uncertainty model. Section 3.5 discusses the evaluation of the integer linear programming models in a Smart-Grid Simulation environment. In Sect. 3.6, we present Conclusions and describe future work.

3.2 Problem Statement

When building a Smart-grid self-healing model, multiple issues need to be included. Some issues pertain to the physical infrastructure, such as the generators, busses, relays, and transmission lines. Other considerations pertain to the cyber infrastructure, such as the communication networks, storage, protocols, security, and procedures for managing the grid. Here, we focus on the issues with the physical infrastructure that involve resource allocation.

3.3 Physical Infrastructure Issues

3.3.1 *Distributed-Device Control Functions*

Most devices associated with nodes in the system must be controllable through remote action. One example is the traditional remote relay-control circuit that is capable of tripping a circuit breaker when the electrical current is higher than the threshold. A second example is adaptive control of inverters to ensure stable voltages. Fully centralized control is impossible, but local-device control with distributed intelligence is highly desirable.

3.3.2 *Selective Load Control*

The ability to selectively switch off customers under certain conditions can help avoid a wide-ranging blackout. This selection process also allows consumers to manage their energy consumption, emphasizing low-cost time periods.

3.4 Micro-Grid Islanding

Distributed Energy Resources (DERs) are small-scale power generators, such as micro-turbines, diesel generators, solar arrays, fuel cells, and wind farms, that are located near a customer cluster. When configured into a micro-grid, these systems automatically disconnect themselves from a single point of connectivity with the primary grid when a disruption occurs. When the primary grid returns to normal conditions, a micro-grid must reconnect and resynchronize its operation.

3.5 Distributed Pathway Control

Alternative, redundant pathways for electricity can be utilized to maintain service under disruptive conditions. The mathematical models that we develop are focused on the distributed pathway control issue, with the objective of finding an optimal set of alternative pathways for electricity to flow from supply sources to demand sites while also satisfying the constraints for the transmission line's capacity [CT99, CLD02, DCN04, DNS95].

3.6 Smart-Grid Modeling

Several models have been developed to characterize the grid functioning under various conditions. A probabilistic model of load-dependent cascading failure is presented in the literature [KJN04, Kru06]. The important area for managing consumers' electric consumption in response to supply conditions and pricing has received attention. The role of factors such as load scheduling and market prices in driving consumer behavior and achieving energy efficiency is described [MWJ10, She95]. User preferences are modeled using the concept of the discomfort level within an optimization-problem formulation that balances the load and minimizes the user's inconvenience that are caused by demand scheduling [She95]. An energy-consumption scheduling problem is established to minimize the overall energy cost [Kad09]. Javed et al. [JAW10] formulated a linear program for distribution management. Kadar [Kad09] developed an optimization model to design the Smart-grid network infrastructure. Our work is the first development of optimization models, specifically for real-time self-healing, that directly incorporate uncertainty. Several studies were done on multi-agent-based architecture in a Smart grid [JW00, NGL11, PFR09, Wan01, NS02].

At the center of any power-system design is the control and communication architecture, which is comprised of the hardware and protocols to exchange critical status and control signals. In conventional electric-power systems, this communication architecture can be achieved with the Supervisory Control and Data Acquisition (SCADA) system [BW03, RI10].

3.7 Integer Linear-Programming Models

Early linear programming (LP) models came into prominence and practice during World War II as a means to improve the efficiency and utilization of scarce resources. LP models have a linear objective function to minimize or maximize as well as linear constraints in the form of inequality equations. Dantzig's [Dan98] simplex method has been a mainstay solution methodology, and the more recent interior-point method is also prominent. Integer ILP models often arise from node-arc network formulations. These network models date to the pioneering work of Ford and Fulkerson [FF10]. Work on the capacitated transshipment problem (CTP) [BBG77] gave the first full descriptions of highly efficient solution algorithms for the type of ILP that applies to the self-healing problem.

In a self-healing Smart grid, we assume that disruptions with the available energy occur due to malfunctioning or failed devices and/or inoperative transmission lines [Ami04, AS08, BCP08, CT99]. These disruptions affect the ability of particular supply sources to meet energy demands at specific sites. In response to the associated need to allocate electrical power in alternative ways in order to accomplish self-healing, we devise several optimization models for increasing complexity to assign supply sources to demand sites. More specifically, we assume that there are J distinct energy demands for which alternative supply sources must be allocated in the short term in order to respond to disruptions. For each of these J demands, there is a finite set of available supply sources that can be allocated to meet the need. We index the supply sources by $i = 1, 2, 3, \ldots, I$. Figure 3.1 shows a bi-partite graph where the supply sources are nodes in the left set and where demand sites are nodes in the right set.

The graph's arcs model intact transmission paths with multiple links that utilize sequences of transmission lines, busses, relays, transformers, capacitors, and other devices. The graph is typically not complete, with missing arcs modeling the

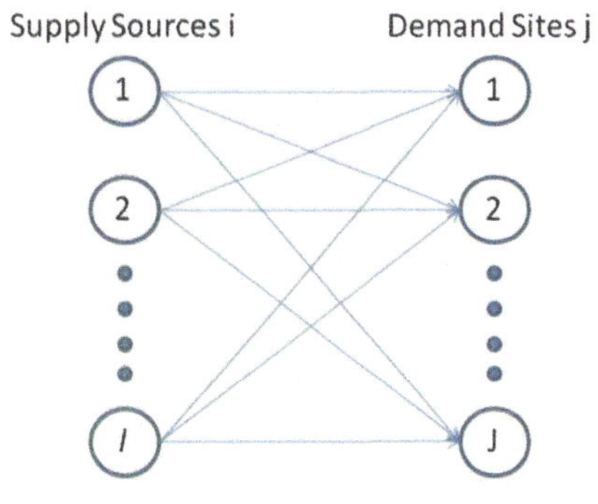

Fig. 3.1 A bi-partite graph with supply and demand sites

unavailability of a viable transmission path. We use c_{ij} to denote the cost of assigning supply source i to demand site j. The objective function's parameters are evaluations of a utility function that includes multiple factors taken together, such as prices established under existing contracts, regulatory principles, prices negotiated in near real time, issues related to the transmission paths' viability, and expected reliability. Supply source i has a specified level, s_i, of energy available; demand sites have a specified level, d_j, of energy needed; sources can supply multiple demand sites; and demand sites can be served from multiple sources. We note that the available supplies and demands can be split freely in their allocations, and the variables, x_{ij}, can be viewed as power flows from supply sources to demand sites. We must also ensure that the transmission paths connecting supply sources to demand sites have sufficient capacity to bear their load levels. In a self-healing situation, we let u_{ij} denote an increased load level (capacity) that can be allocated to an available pathway connecting nodes i and j. This node-to-pathway relationship leads to the following problem:

$$\text{Max } z = \sum_{i=1}^{I} \sum_{j=1}^{J} c_{ij} x_{ij} \tag{3.1}$$

Subject to

$$\sum_{i=1}^{I} x_{ij} \geq d_j \text{ for all } j \tag{3.2}$$

$$\sum_{j=1}^{J} x_{ij} \leq s_i \text{ for all } i \tag{3.3}$$

$$0 \leq x_{ij} \leq u_{ij} \tag{3.4}$$

One limitation of this basic model is the implicit assumption that the transmission paths modeled by the arcs have no links in common, which may not be the case in practice. This assumption leads to an expanded model formulation that breaks the bipartite graph into a more general network and includes capacities for the individual links:

3.8 Notation

The directed graph (network) has node set N and link set $A = N \times N$. We denote typical elements:

i belongs to N, and (i,j) belongs to A.
c_{ij} = utility per power flow unit on (i,j).
u_{ij} = capacity (upper bound) of (i,j).

b_i = supply of power at node i (interpret negative b_i as a demand of -b_i units).
Variable x_{ij} = power flow on link (ij).

The problem is to find the set of flows that minimizes the total cost subject to constraints which require (i) "flow balance" at each node and (ii) capacity restriction for each link. The formulation is as follows:

$$\text{Max } z = \sum_{(i,j) \in A} c_{ij} x_{ij}$$

Subject to.

$$\sum_{i:(i,j) \in A} x_{ij} - \sum_{j:(i,j) \in A} x_{ji} = b_i \text{ for all } i\, j \tag{3.5}$$

$$0 \leq x_{ij} \leq u_{ij} \text{ for all } i, j \tag{3.6}$$

Constraint set 1 consists of flow-balance constraints. The first term in such a constraint is summed for all links with a tail at node i, which is referred to as the forward star of node i. Similarly, the second term is summed for all links with a head at node i, the "reverse star" of node i. This model requires that the total supply and total demand are equal. The model is known as the Capacitated Transshipment Problem (CTP) in the literature. Figure 3.2 illustrates the topology for this type of network.

More generally, it may be important to explicitly distinguish supply sources by type. For example, if a site supplies power with wind, there may be specific, important information about that source, such as uncertainty. In the following model, supply sources and demand sites are indexed and differentiated by type, p, where the index takes on values p = 1, 2, ..., P. Accordingly, we now have the following notation:

Parameters:

c_{ijp} = utility per unit of flow of type p on link (i,j).
u_{ij} = capacity (upper bound) for flow on link (i,j).
b_{ip} = supply of power of type p at node i (interpret negative b_i as a demand of -b_i).
x_{ijp} = flow of power of type p on link (ij).

$$\text{Max } z = \sum_{(i,j) \in A} \sum_{p \in P} c_{ijp} x_{ijp}$$

Subject to

$$\sum_{i:(i,j) \in A} x_{ijp} - \sum_{j:(i,j) \in A} x_{jip=} = b_{ip} \quad \text{For all } (i, j) \tag{3.7}$$

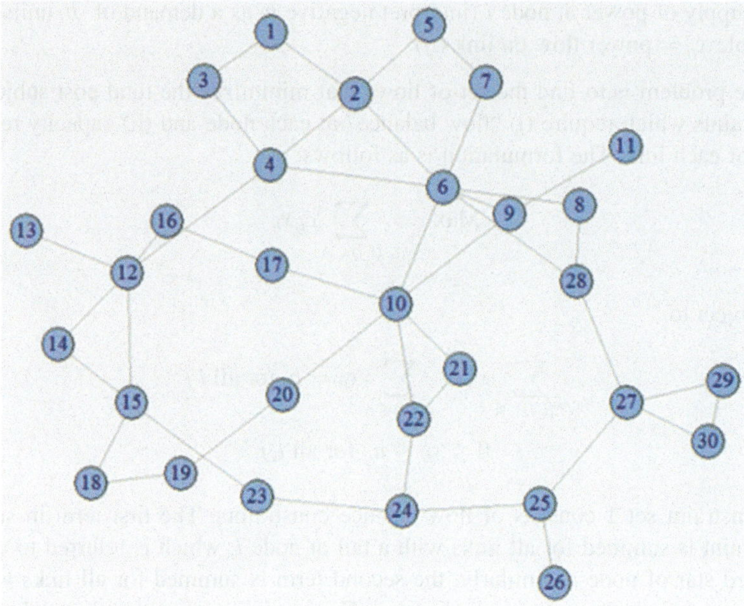

Fig. 3.2 Smart-grid topology

$$\sum_{p=1}^{P} x_{ijp} \leq u_{ij} \qquad \text{For all } (i,j) \tag{3.8}$$

$$x_{ijp} \geq 0 \text{ For all } (i,j) \text{ and } p \tag{3.9}$$

In the literature, this type of modeling is known as the multi-commodity CTP. The first constraint set enforces that flow balance must occur for each type of power through every node i. The value of b_{ip} is positive at strictly supply source nodes, negative at strictly demand-site nodes, and zero at pure transshipment nodes. The model allows for supply sources or demand sites to also serve as transshipment points, but such a transshipment arrangement would be unusual in practice. The second constraint set allows for each link in the distribution system to be restricted by joint capacity for all flows that pass through it. The model is NP-complete.

3.9 Uncertainty in Resource Allocation

We now consider the possibility that supplies and demands at certain nodes are uncertain, as is often the case for supply sources such as wind or solar power. The typical power curve shown in Fig. 3.3 illustrates the uncertainty of the power output that can be obtained from a wind machine.

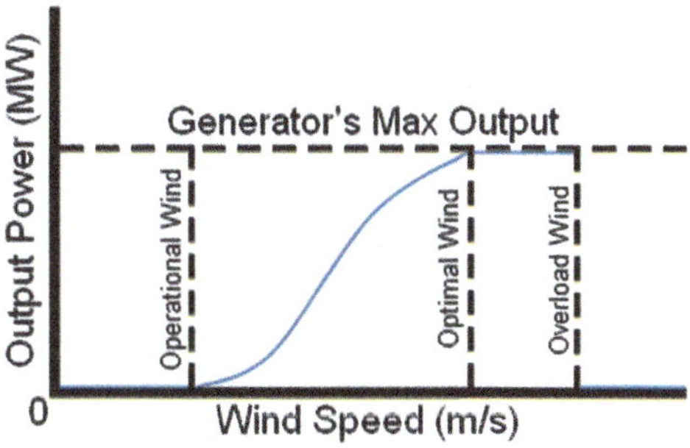

Fig. 3.3 Wind-machine power curve

For a given source node i and power type p, we modify a constraint in set (1) to make it probabilistic as follows:For a specific i

$$\Pr\left[\sum_{i:(i,j)\in A} x_{ijp} = b_{ip}\right] \geq 1 - \alpha_{ip} \tag{3.10}$$

For an easier explanation, we assume that node i is the sole source of commodity type p and that it does not serve as a transshipment point for power originating at other sites. In this constraint, $1 - \alpha_{ip}$ is the pre-assigned, smallest-allowable probability with which the power available from source i is sufficient to supply b_{ip} units for a demand site. We view α_{ip} as the acceptable risk of not receiving b_{ip} MegaWatts [MW] of electrical power from the specific DER source. For specific values of i and p, we assume that b_{ip} is a random variable that follows a statistical distribution. We note that varying the value of b_{ip} results in different flows through the network links which then, in turn, affects the links' capacity constraints. In the case where b_{ip} follows the normal distribution with mean $E\{b_{ip}\}$ and variance var. $\{b_{ip}\}$, we standardize the random variable by subtracting the mean and dividing by the square root of the variance, resulting in the following equivalent, probabilistic condition:

$$\Pr\left[\left[\frac{\sum_{i:(i,j)\in A} x_{ijp} - E\{b_{ip}\}}{\sqrt{\mathrm{Var}\{b_{ip}\}}}\right] = \left[\frac{b_{ip} - E\{b_{ip}\}}{\sqrt{\mathrm{Var}\{b_{ip}\}}}\right]\right] \geq 1 - \alpha_{ip} \tag{3.11}$$

The true meaning of the equation in the application should be to enforce the condition that the power distributed from supply source i to its outgoing links is at a level for which there is confidence that at least that much power will actually be delivered with a prescribed probability. Any overage would likely be dissipated. This consideration makes it legitimate to replace the equation with an inequality in the analysis:

$$\Pr\left[\left[\frac{\sum_{i:(i,j)\in A} x_{ijp} - E\{b_{ip}\}}{\sqrt{\mathrm{Var}\{b_{ip}\}}}\right] \geq \left[\frac{b_{ip} - E\{b_{ip}\}}{\sqrt{\mathrm{Var}\{b_{ip}\}}}\right]\right] \geq 1 - \alpha_{ip} \tag{3.12}$$

We let Φ represent the cumulative distribution function for the standard normal distribution and let $K_{\propto_{ip}}$ be the standard normal value such that

$$\Phi(K_{\propto_{ip}}) = 1 - \propto_{ip} \tag{3.13}$$

Then, the probabilistic condition given above is realized if

$$\left[\frac{\sum_{i:(i,j)\in A} x_{ijp} - E\{b_{ip}\}}{\sqrt{\mathrm{Var}\{b_{ip}\}}}\right] \geq K_{\propto_{ip}} \tag{3.14}$$

This equation can be rewritten as a constraint:

$$\sum\nolimits_{i:(i,j)\in A} x_{ijp} \leq E\{b_{ip}\} + K_{\propto_{ip}}\sqrt{\mathrm{Var}\{b_{ip}\}} \tag{3.15}$$

This constraint gives the condition that the power delivered will be within the upper-bound value given by the right-hand side with a probability $1 - \alpha_{ip}$. By the symmetry of the normal distribution, constraint

$$\sum\nolimits_{i:(i,j)\in A} x_{ijp} \leq E\{b_{ip}\} - K_{\propto_{ip}}\sqrt{\mathrm{Var}\{b_{ip}\}} \tag{3.16}$$

sets the requirement for the minimum level of power that is delivered with the prescribed probability. This equation is a linear constraint that is incorporated into the optimization problem as a so-called "chance constraint," effectively modeling probabilistic conditions within a linear program. As an example, suppose that the supply for node 3 is a wind source that provides power with a mean value of 7 MW and a variance of 4 MW, and that the supply has outgoing distribution links to transshipment nodes 4, 7, and 8. Node index 3 also identifies the type of power generated at node 3. If we allow a 5% risk for not meeting the supply objective, we have the following condition:

$$x_{343} + x_{373} + x_{383} \leq 7 + 1.645 \times 2$$

or

$$x_{343} + x_{373} + x_{383} \leq 11.935 \tag{3.17}$$

The value 1.645 comes from a table of standard normal variates. The condition means that there is a 95% chance that the realizable power from the wind source is no more than 11.935 MW. Using the symmetry,

$$x_{343} + x_{373} + x_{383} \leq 7 - 1.645 \times 2$$

or

$$x_{343} + x_{373} + x_{383} \leq 3.71 \qquad (3.18)$$

This constraint means that at least 3.71 MW of power can be realized with 95% probability. If we increase the prescribed probability to a more stringent 99%, the standard normal variate value is 2.33, and the constraint becomes

$$x_{343} + x_{373} + x_{383} \leq 2.34 \qquad (3.19)$$

The model can also be readily extended to multiple indexed time periods with a time-dependent, supply-demand allocation that has fixed costs. This formulation is important for consistency with the time-period planning granularity models used by most utilities.

3.10 Smart-Grid Simulation

Our Smart-grid simulator runs as a Multi-Agent System (MAS) using the Java Agent Development Framework (JADE). Software agents act autonomously and communicate with each other across open and distributed environments, making an agent design ideal for simulating a Smart grid. The agents can sense, act, communicate, and collaborate with each other; be empowered with degrees of autonomy; are decentralized; and have local views and knowledge. The simulation has a low-level, physical-device layer with components that can exhibit fault conditions and fail. A middle layer has agents with a knowledge base, including consumer, DER, device-managing, and monitoring agents. An upper layer consists of management agents that receive the system's state information, carry out analyses, and invoke decision-support models. The optimization models described in this chapter reside at this third level. However, the simulator is also designed to support suites for decision-support models, including fuzzy logic, statistical hypothesis testing, Bayesian belief networks, and constraint satisfaction. These agents also stream reporting information, allowing for convenient performance comparisons [GK03, FG96, KH09, Mar94].

When a three-layer optimization model generates a workable solution in a self-healing situation, solution values are then converted into the associated corrective actions that are done at lower layers to invoke the appropriate response. Each corrective action is modeled by an agent-task-pair. The task breaks into detailed roles and actions at the device and transmission-line level. A graphical user interface allows human intervention, if appropriate, or autonomous execution by simply setting initial values for parameters, conditions, and state information.

3.11 Conclusions

The developed optimization models include objective functions that maximize a utility function and constraints that ensure feasibility for the resource allocation. Stochastic information can be directly included with the constraints to model situations with known uncertainty. The agent-based simulation provides a realistic and readily validated means to evaluate the performance of the integer linear-programming solutions as they would function in an operational Smart grid.

Chapter 4
Resource Allocation Using Branch and Bound

The chapter describes a resource-allocation problem in a smart-grid application that is formulated and solved as a binary integer-programming model. To handle power outages from the main distribution circuit, the Smart grid's intelligent agents have to utilize and negotiate with distributed-energy resource agents that act on behalf of the grid's local generators in order to negotiate power-supply purchases to satisfy shortages. We develop a model that can optimally assign these DERs to the available multiple regional utility areas (RUAs) or units that are experiencing power shortages. This type of allocation is a resource-assignment problem. The DERs in our model depict the behavior of power created with a wind turbine, solar generation, or other renewable generation units, and the region or area refers to a centralized distribution unit. The integer-programming approach is called Capacity-Based Iterative Binary Integer Linear Programming (C-IBILP). All simulation results are computed using the optimization tool box in MATLAB. Computation results exhibit very good performance for the problem instances tested and validate the assumptions made.

The material in this chapter was co-authored by Prakash Ranganathan and Kendall Nygard. Prakash Ranganathan had primary responsibility for developing the linear-programming formulation of a resource-allocation problem using the branch-and-bound method. Prakash Ranganathan was the primary developer for modeling, implementing, and testing the conclusions that are advanced here. Prakash Ranganathan also drafted and revised all versions of this chapter. Kendall Nygard served as a proofreader and checked the LP formulation that was run by Prakash Ranganathan.

P. Ranganathan, K.E. Nygard, *Distributed Linear Programming Models in a Smart Grid*, Power Electronics and Power Systems, DOI 10.1007/978-3-319-52617-1_4

4.1 Distributed Energy Resources in a Smart Grid

Dynamic, real-time power systems often operate in continuously changing environments, such as adverse weather conditions, sudden transformer failures, or a malfunctioning sub-system in a transmission or distribution network. These disruptions, along with the complexity of power systems, cause the power system's energy demands and loads to fluctuate, potentially resulting in widespread outages and huge price spikes. Data from the North American Electric Reliability Council (NERC) and analyses from the Electric Power Research Institute (EPRI) indicate that the average outages from 1984 to 2004 affected nearly 700,000 customers annually [Ami04]. Smaller outages occur much more frequently and affect tens to hundreds of thousands of customers every few weeks or months while larger outages occur every 2–9 years and affect millions. Although preventing these outages remains a challenge, such demand changes (increases or decreases) from consumers can often be offset by distributed energy resources (DERs) which are renewable resources. Solar- and wind-based power can satisfy the shortages or reduce the outage levels. In our work, we consider the use of such DER-based standby mechanisms to support any additional demand. We apply an Iterative Binary Integer Linear Programming (IBILP) technique [Web01] to optimally assign DERs for a region based on criteria such as power levels, demands, and preferences. Resource allocation for a complex power system is robust with respect to demand variations and power-level fluctuations. The amount of additional power that DERs can generate and that can be effectively utilized in a power network is a measure of robustness. Hence, we argue that a capacity-based Iterative Binary Integer Linear Programming (C-IBILP) model is, inherently, a robust resource allocation.

The structure for the rest of the chapter is as follows. In Sect. 4.2, an overview and Related Work for the smart grid is discussed. In Sect. 4.3, we present a general formulation of this DER assignment problem. In Sect. 4.4, we describe how to solve this problem optimally by using a branch-and-bound based (BB) algorithm with equality and inequality constraints. In Sect. 4.5, we show the experimental results.

4.2 Related Work

Mathematical programming has enjoyed a burgeoning presence in theoretical computer science, both as a framework for developing algorithms and, increasingly, as a bona fide model of computation where the limits are expressed in terms of the formulations' sizes and the formulations' integrality gaps [ABL06, AAT05, BJN98]. Linear formulations are an appealing model for computation because both optimization and decision problems fit naturally into the framework and because both theoretically tractable and efficient practical algorithms exist to solve linear programs. For instance, state-of-the-art approaches to exactly solve large-scale instances of many NP hard problems rely on integer-programming approaches that require the repeated solution of integer programs that represent the problems [BJN98]. The polynomial-time algorithms [Kru56] and other algorithms [KY97, KP94] cannot be used in this application due to high complexity and extensive run-times. Modification will be investigated in future work. We refer to a fundamental model for DER assignment as the capacity-based Iterative Binary Integer Linear Programming (C-IBILP) model. There has been little attention given to this approach in smart electrical-grid analyses. To our knowledge, these smart-grid problems have not been solved for DER allocations using models that perform optimal matching for supply sources' demand sites by predicting generation and market-controlled consumption. Such optimization algorithms are comparable to hard, unsolved problems about inference, optimization, and control [Web02].

4.3 Assigning DER to RUA Formulation

To illustrate our problem formulation, we assume that there are seven areas (RUAs) and six DER units with the demand and preference levels shown in Figs. 4.1 and 4.2 respectively. We define a regional utility area (RUA) as the local-distribution power utilities within the micro-grid that distribute power within their network for its loads [Ami04]. For simplicity, we name them Area1, Area 2, . . ., Area 7 as illustrated in Fig. 4.1. The power demand and the preferences in Fig. 4.2 depict a demand-driven DER assignment problem that accommodates preference information. The parameters in the figure are for illustration purposes.

A simple allocation "text" script in MATLAB would be as follows: text (0.1, .73, 'area1'); text (.35, .73, 'area2'); text (.60, .73, 'area3'); text (.82, .73, 'area4'); text (.35, .42, 'area5'); text (.60, .42, 'area6'); text (.82, .42, 'area 7').

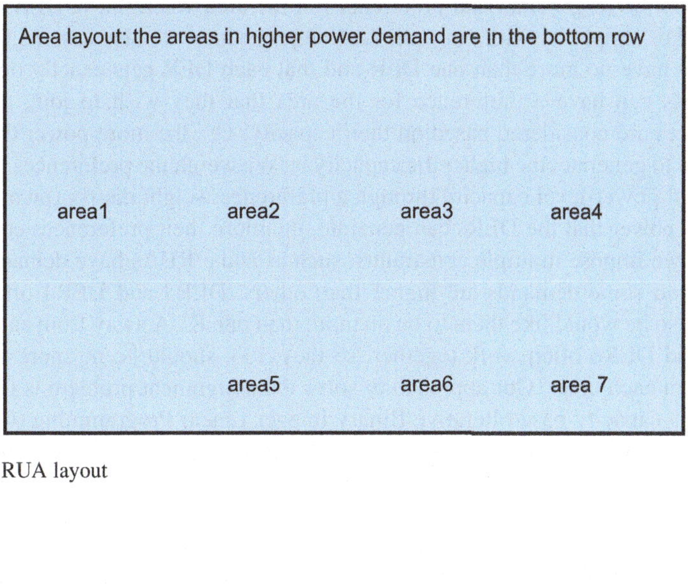

Fig. 4.1 RUA layout

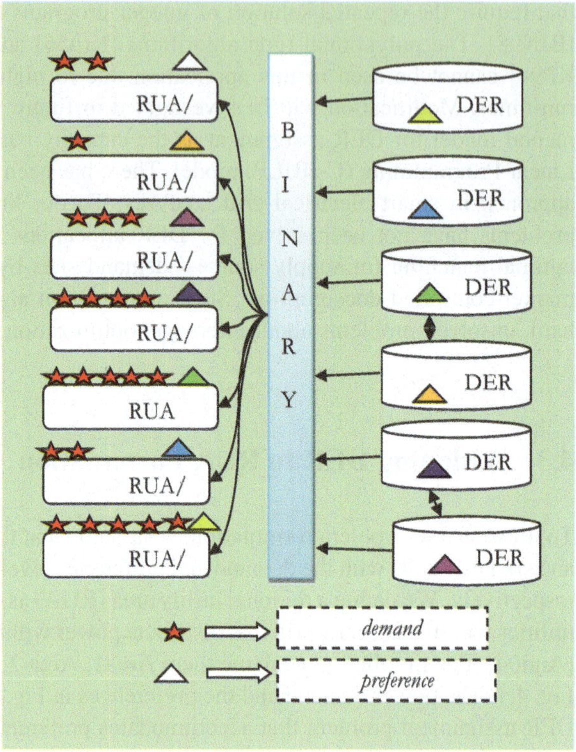

Fig. 4.2 DER vs. RUA assignment problem

For example, suppose our simulation study is charged with a need to optimally assign six DERs, DER1, DER2, DER3, DER4, DER5, and DER6, to seven regional utility areas (RUAs) based on criteria such as the power-level capacity that these DERs are able to generate and preferences in the area where these DERs wish to operate. For simplicity with our optimization procedure, we also assume that each RUA can have no more than one DER and that each DER gets exactly one RUA. The DERs can have a preference for the area that they wish to join, and their preferences are considered based on their capacity; i.e., the more power they have been able to generate, the higher the capacity is. We weigh the preferences based on the DERs' power-level capacity through a preference weight matrix (pwm) so that the more power that the DERs can generate, the more their preferences count.

Also, we impose multiple constraints, such as some RUAs have demand, some do not, and some demands are higher than others. DER3 and DER4 often work together, so be would like them to be no more than one RUA away from each other; DER5 and DER6 often work together, so they, too, should be no more than one RUA from each other. Our approach to solve the assignment problem is to formulate it as a capacity-based Iterative Binary Integer Linear Programming (C-IBILP) model and to relax the integrality constraints. Our overall objective is to maximize the satisfaction for the preferences weighted by capacity which will allocate these DERs to their areas. This task is done through a binary integer-programming model by defining a linear objective function. Our algorithm uses a branch-and-bound procedure with linear-programming bounds that have "minimum integer infeasibility" as the branch strategy and a "depth-first search" for the node-search strategy.

To develop our problem formulation, the first step is to choose what each element of the solution vector, |*x*|, represents. We use binary integer variables which represent the specific assignments of DERS to RUAs. If the DER is assigned to a RUA, the variable takes the value 1, and if not assigned, the variable takes the value 0. We consider the DERs in sequential order as DER1, DER2, DER3, DER4, DER5, DER6, and DER7. The *n*th sequence of elements in vector |x| stores the assignment variables for DER n. In our example, |x (1)| to |x (7)| corresponds to DER1 being assigned to Area 1, Area 2, etc., up to Area 7. In all, vector |x| has 6 sequences of 7 elements each, or 42 elements in all. Each sequence has a single binary variable set to 1, enforcing a multiplc-choicc condition for each DER.

4.4 DER Capacities

We impose constraints based upon the DER preference level in the area of operation driven by the capability to generate power. The concept is that, the more power a DER can generate, the higher the preference level is. For example, consider the following randomly set power levels given in kilowatts (kW).

DER1→9 kW.
DER2→10 kW.
DER3→5 kW.
DER4→3 kW.
DER5→1.5 kW and
DER6→ 2 kW.

We create a normalized weight vector based on capacity and also assume that certain DERs should be used in some preferred region or area, such as a DER with more power-generation capability being used in high-demand areas. This normalized weight vector can be obtained in MATLAB as follows:

capacity = [9 10 5 3 1.5 2];

weight vector = capacity/sum (capacity);

```
>> capacity

capacity =

    9.0000   10.0000    5.0000    3.0000    1.5000    2.0000

>> weightvector

weightvector =

    0.2951    0.3279    0.1639    0.0984    0.0492    0.0656
```

4.5 RUA Preferences

We set up a preference weight matrix (pwm or prefmatrix) where the rows correspond to areas and the columns correspond to DERS. We assume that each DER will give values for each area so that the sum of all their choices (i.e., their columns) is 100. A higher number means that the DER prefers the area. We justify using the preference matrix by noting that limitations in algorithm scalability and data availability preclude a fully centralized solution for the problem of interest. Thus, decision making must be decentralized, and we, accordingly, divide the power network into many smaller RUAs, where the prefmatrix concept is applied to individual regions in the network. An example of DER preferences is shown here:

DER1 = [0; 0; 0; 0; 10; 40; 50];
DER2 = [0; 0; 0; 0; 20; 40; 40];
DER3 = [0; 0; 0; 0; 30; 40; 30];
DER4 = [1; 3; 3; 3; 10; 40; 40];
DER5 = [3; 4; 1; 2; 10; 40; 40];
DER6 = [10; 10; 10; 10; 20; 20; 20];

The *i*th element of a DER's preference vector is the value of the *i*th RUA. Thus, the combined preference matrix is expressed as "prefmatrix":prefmatrix = [DER1 DER2 DER3 DER4 DER5 DER6];

```
>> prefmatrix

prefmatrix =

     0     0     0     1     3    10
     0     0     0     3     4    10
     0     0     0     3     1    10
     0     0     0     3     2    10
    10    20    30    10    10    20
    40    40    40    40    40    20
    50    40    30    40    40    20
```

4.5.1 *Case 1*

We treat the above "prefmatrix" arrangement as case 1 for analysis. We then weigh the preference matrix with the |weight vector| to scale the columns by capacity. We also reshape this matrix as a vector in column-order so that it corresponds to the |x| vector. This task is achieved in MATLAB script as follows:

PM = prefmatrix * diag (weightvector);

```
>> diag(weightvector)

ans =

    0.2951         0         0         0         0         0
         0    0.3279         0         0         0         0
         0         0    0.1639         0         0         0
         0         0         0    0.0984         0         0
         0         0         0         0    0.0492         0
         0         0         0         0         0    0.0656
```

c = PM (:);

```
>> PM

PM =

         0         0         0    0.0984    0.1475    0.6557
         0         0         0    0.2951    0.1967    0.6557
         0         0         0    0.2951    0.0492    0.6557
         0         0         0    0.2951    0.0984    0.6557
    2.9508    6.5574    4.9180    0.9836    0.4918    1.3115
   11.8033   13.1148    6.5574    3.9344    1.9672    1.3115
   14.7541   13.1148    4.9180    3.9344    1.9672    1.3115
```

The objective is to maximize the total preference measure weighted by capacity. This is a linear objective function, max c'*x or, equivalently, min -c'*x, with c being the DER preferences. We use the BINTPROG script of MATLAB to run our model that is defined as follows:

$$\min_{x} f^T x : \begin{cases} A.x \leq b, \\ \text{Aeq}.x = \text{beq}, \\ x : \text{binary} \end{cases}$$

where

f = Vector containing the coefficients of the linear-objective function.A = Matrix containing the coefficients of the linear-inequality constraints, $A \cdot x \leq b.b$ = Vector corresponding to the right-hand side of the linear-inequality constraints. Aeq = Matrix containing the coefficients of the linear-equality constraints, $Aeq \cdot x$ = beq.beq = Vector containing the constants of the linear-equality constraints.$x0$ = Initial point for the algorithm.Options: Option structure containing the algorithm's options.x: A binary integer solution vector: that is, its entries can only take the values 0 or 1.

4.6 Constraints

The first set of constraints requires that each DER is assigned to exactly one area. For example, because DER2 is the second DER, we enforce the condition that lsum (x (8:14)) = 1l. We represent these linear constraints in an equality matrix, Aeq, and right-hand side vector, beq, where lAeq*x = beql, by building the appropriate matrices. Matrix lAeql consists of ones and zeros. For example, the second row of lAeql corresponds to DER2 getting exactly one RUA, so the row pattern is as follows:

0 0 0 0 0 0 0 1 1 1 1 1 1 1 0 0 0 0 0 0 0 0 0 0 0 0 0 ... 0 0 0

These conditions are implemented in MATLAB code as follows:

|Aeq (2,:)*x = 1| is equivalent to |sum(x (8:14)) = 1|.

numAREAS = 7;
numDERS = 6;
numDim = numAREAS * numDERS;
onesvector = ones (1, numAREAS);
Each row of Aeq corresponds to one DER.
Aeq = blkdiag (onesvector, onesvector, onesvector, onesvector, onesvector, onesvector);
beq = ones (numDERS, 1);
view the structure of Aeq, that is, where there are nonzeros (ones) Figure;

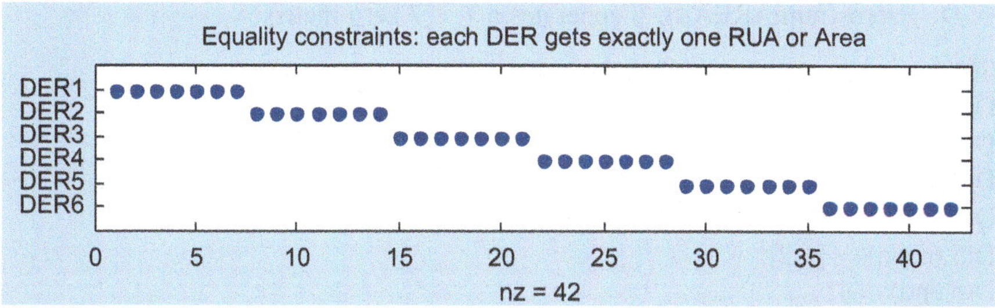

Fig. 4.3 Equality constraints

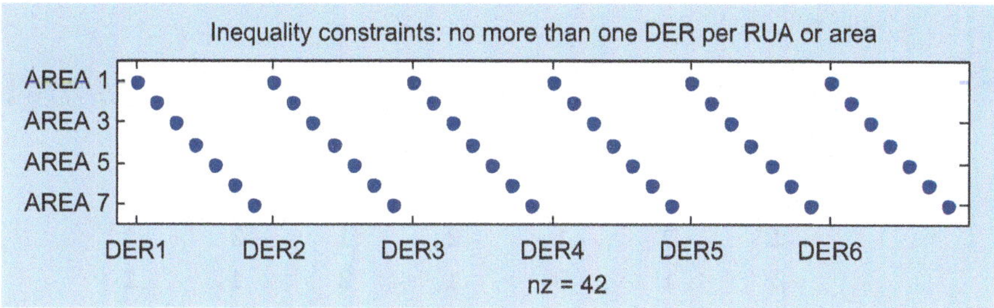

Fig. 4.4 Inequality constraints

The second sets of constraints are inequalities. These constraints specify that each area has no more than one DER in it; i.e., each AREA has one DER in it or is empty. We build matrix |A| and vector |b| such that $|A^* \leq b|$ to capture these constraints. Each row of |A| and |b| corresponds to a RUA, so row 1 corresponds to the DER assigned to RUA 1. In this case, the rows have the pattern type shown below for row 1:

1 0 0 0 0 0 0 1 0 0 0 0 0 0 1 0 0 0 0 0 0 ... 1 0 0 0 0 0 0

Each subsequent row is similar but is shifted (circularly) to the right by one spot. For example, row 3 corresponds to RUA 3 and enforces $|A(3,:)^*x \Leftarrow 1|$ so that AREA 3 cannot have more than one DER. Figures 4.3 and 4.4 illustrate the equality and inequality constraints.

A = repmat (eye (numAREAS),1, numDERS);
b = ones (numAREAS,1);

where repmat represents the replicate and tile array. Elements of the next constraint set are also inequalities, so they are added to matrix |A| and vector |b| that already contain the previous inequalities. We wish to enforce that DER3 and DER4 are no more than one area (RUA) from each other, and similarly, DER5 and DER6 are no more than one area away from each other. First, the symmetric distance matrix for the RUAs is built using physical locations and Manhattan (i.e., the "taxicab" metric).

D = zeros(numAREAS); // generates a 7 × 7 zero matrix.

Setting up the top-right half of the matrix,
D(1,2:end) = [1 2 3 2 3 4];
D(2,3:end) = [1 2 1 2 3];
D(3,4:end) = [1 2 1 2];
D(4,5:end) = [3 2 1];
D(5,6:end) = [1 2];
D(6,end) = 1;

The lower-left half is the same as the upper-right D = triu(D)' + D. We find the RUAs that are more than one distance unit away.

```
>> D

D =

     0     1     2     3     2     3     4
     1     0     1     2     1     2     3
     2     1     0     1     2     1     2
     3     2     1     0     3     2     1
     2     1     2     3     0     1     2
     3     2     1     2     1     0     1
     4     3     2     1     2     1     0

>> D = triu(D)' + D;
>> D

D =

     0     1     2     3     2     3     4
     2     0     1     2     1     2     3
     4     2     0     1     2     1     2
     6     4     2     0     3     2     1
     4     2     4     6     0     1     2
     6     4     2     4     2     0     1
     8     6     4     2     4     2     0
```

[AREAA,AREAB] = find(D > 1);
numPairs = length(AREAA);

This finds |numPairs| area pairs. For example, if DER3 occupies one area in the pair, then DER4 cannot occupy the other AREA in the pair; otherwise, it would be more than one unit away in terms of AREA. The same condition holds for DER5 and DER6. This situation gives |2*numPairs| additional inequality constraints which we add to |A| and |b|. By adding rows to A, we accommodate these constraints as follows:

numrows = 2*numPairs + numAREAS;

A((numAREAS+1):numrows, 1:numDim) = zeros(2*numPairs,numDim);

For each pair of AREAS in numPairs, for the |x(i)| that corresponds to DER3 in |AREAA| and for the |x(j)| that corresponds to DER 4 in |AREAB|, $x(i) + x(j) \leq 1$; i.e., either DER3 or DER4, but not both, can occupy one of these AREAS.

4.7 Branch-and-Bound (BB) Strategy

The branch-and-bound algorithm is a well-known, optimal solution method. Branch-and-bound (BB) algorithms are methods for solving non-convex global-optimization problems [BB91, LW66, BJN98, Moo91]. They are exact (non-heuristic) in the sense that they calculate a provable upper and lower bound on the globally optimal objective value and that they terminate when all suboptimal feasible solutions have been eliminated. Branch-and-bound algorithms can be computationally slow. In the worst case, they require effort that grows exponentially with problem size. We achieve fast convergence in our problems. We note that, due to total unimodularity of the basic A matrix, a network-based, customized linear-programming solver could be used to provide the lower bounds very quickly in large problems. The BB algorithm is a well-known algorithm in the research community [Wol98]. An example run of the BB algorithm is shown in Fig. 4.5, followed by a snippet of MATLAB code showing the iterative output for each node displayed in the branch-and-bound algorithm. We let the BINTPROG choose the starting point.

```
x0 = [];
f = −c;
options = optimset('Display','iter','NodeDisplayInterval',1);
[x,fval,exitflag,output] = bintprog(f,A,b,Aeq,beq,x0,options);
fval
exitflag.
output.
```

To reduce the number of nodes explored, the time, or the number of iterations taken, there are alternative options available. BINTPROG uses the options to adjust the algorithm with differing nodes and branching-variable strategies [Moo91, MG05].

For example, the default branching strategy is |'maxinfeas'|, which chooses the variable with the maximum integer infeasibility for the next branch, that is, the

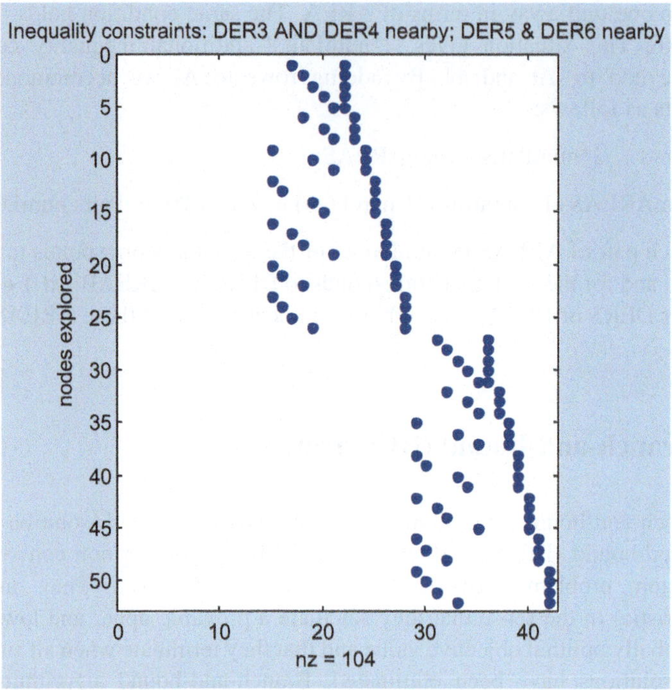

Fig. 4.5 Branch-and-bound algorithm with inequality constraints

variable with the value closest to 0.5. Running the problem again with the branching strategy set to |'mininfeas'|, the option of minimum integer infeasibility is chosen (that is, the variable where the value is closest, but not equal, to 0 or 1.

To structure the tree, depth-first and best-node search strategies are available. For example, in "df," at each node in the search tree, if there is a child node one level down in the tree that has not already been explored, the algorithm chooses one such child to search. Otherwise, the algorithm moves to the node one level up in the tree and chooses a child node one level down from that node. In a best-node (bn) strategy, the node with the lowest bound on the objective function is the default. In our limited computational experience, convincing and acceptable results are quickly reached. For future work, we plan to increase the scale of our test problems and to investigate improved BB schemes.

4.8 Results

The simulation is done with a MATLAB platform. The prebuilt in command for branch-and-bound algorithm was used to simulate the following cases.

4.8.1 Case 1

The results show that the optimal value is reached in 1.22 s after 163 iterations with 54 nodes participating (case 1) using the capacity-based Iterative Binary Integer Linear Programming (C-IBILP)-based branch-and-bound method which maximizes the satisfaction of the DER preferences weighted by area capacities. The final output shown in Fig. 4.6 presents the DER allocation with RUA1, or area 1, treated as empty for optimal assignment.

4.8.2 Case 2

If we change the DER preferences according to the matrix shown below, then the optimal solution is reached in 0.047 s with 13 iterations and 1 node by using the default-node and branch strategies shown in Fig. 4.7.

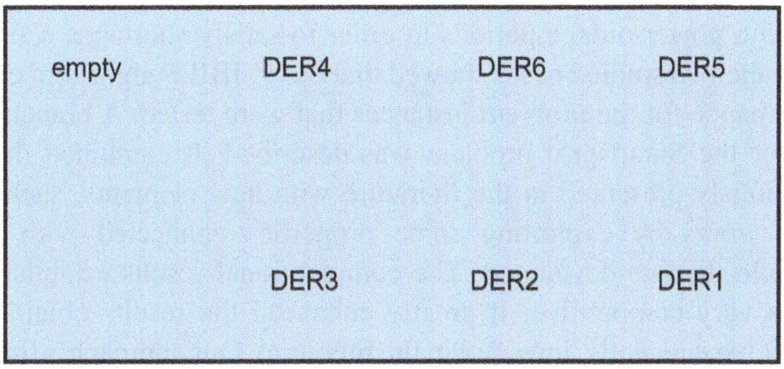

Fig. 4.6 An optimal DER assignment solution for case 1

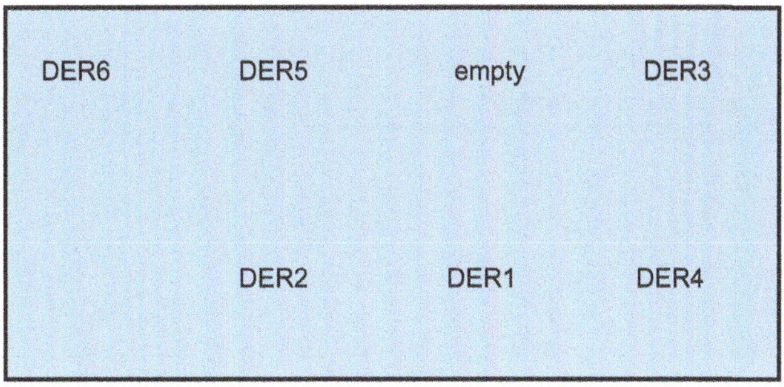

Fig. 4.7 An optimal DER-assignment solution for case 2

```
>> prefmatrix

prefmatrix =

     0     0     0     1     3    70
     0     0     0     3    40    10
     0     0     0     3     1    10
     0     0    70     3     2    10
    10    90    30    10    10    20
    40    40    40    60    40    20
    50    40    30    40    40    20
```

4.9 Conclusions

The chapter presented a resource-assignment problem for Smart-grid applications. The Capacity-Based Iterative Binary Integer Linear Programming (C-IBILP) model was designed to specify an optimal allocation of distributed energy resources (DERs) during power-outage periods in order to satisfy shortages. Computational results from the two studied cases showed that our C-IBILP algorithm exhibits very good performance for the problem instances that were tested. A branch-and-bound algorithm for the Smart-grid problem was described. It combined the extension results previously presented in the literature with new elements, such as a lower bound that works by exploiting some properties connected with the ad-hoc branching rule that we developed. The computational results established that the algorithm is very competitive. It greatly enhanced the results obtained with the methods that have recently appeared in the literature. Our approach's limitation was that the method does not scale well for larger DERs. Our current efforts involve extending this assignment model to a more scalable assignment formulation where more DERs can serve each RUA.

Chapter 5
Resource Allocation Using DW Decomposition

Dantzig-Wolfe decomposition is a technique for dealing with linear- and integer-programming problems that have embedded substructures that permit efficient solution. The technique has been successfully applied in a variety of contexts [Dan63, Chv83, BJN98]. Implementing DW-decomposition-based algorithms poses various challenges. The primary constraint revolves around convergence of the dual-bound computations and, in the context of integer programming, the enforcement of integrality restrictions. The standard view of DW decomposition is that it exploits the linear-programming formulation of the Lagrangian dual. This so-called master linear program has an exponential number of variables that are handled using dynamic column generation. An alternative view is that DW is a reformulation technique that gives rise to a mixed-integer master program. Viewing DW as a reformulation technique allows for the development of a theoretical framework that facilitates the handling of branching decisions and cutting planes in the master program.

5.1 Why Decompose?

There are computational and organizational advantages when using decomposition algorithms. From a computational perspective, the advantage is that sub-problems are usually easier to solve than the original problem. The sub-problems are, by definition, smaller than the original problem. In many cases, the sub-problems have special properties, such as convexity, sparsity, or network constraints, that enable the use of efficient, specialized algorithms to solve them [Dan63]. By decomposing the original problem, we can take advantage of the efficient solution method available for sub-problems. Decomposition algorithms also allow Smart-grid problems to be solved in a distributed manner. The key point when designing a decomposition algorithm in this environment is that only limited communication between the sub-problems and the master problem is required. The aim is that

P. Ranganathan, K.E. Nygard, *Distributed Linear Programming Models in a Smart Grid*, Power Electronics and Power Systems, DOI 10.1007/978-3-319-52617-1_5

different engineering teams or subgroups of a Smart grid, such as the transmission side or distribution side, can solve only their own sub-problems and that only a small amount of communication is required with the central coordinator.

5.2 Objective Function and Illustration of the DW Algorithm

The linear-programming problem set up for the Dantzig-Wolfe solution technique can be formulated as follows:

Minimize Cx.

$Ax = b$ (Master problem constraints).

$x\in$ to X (where X is a set of corner points).

To illustrate the Dantzig-Wolfe decomposition method, we first consider a 4-bus system with 2 generators and 3 loads that plans to maximize its power based on certain constraints (Fig. 5.1). A bus is a communication link that transports energy from one point to another point. A network model of the 4-bus system is shown in Fig. 5.2. The buses are declared using variables $x1$, $x2$, $x3$, and $x4$. The objective is to maximize the power flow while keeping these constraints in mind. Constraints 1 and 3, namely C1 and C3, describe the line voltages that these buses should not exceed: 5 kv and 8 kv, respectively. Constraints C2 and C4 specify the average repair time needed for these buses to correct themselves in the event of any failure, and the time should not exceed 12 h/year and 10 h/year, respectively. Constraint C5 tells us that the total bus reactive load power should not exceed 7 kW, and

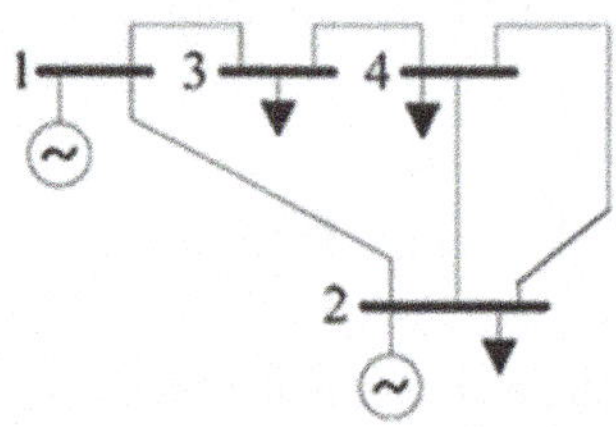

Fig. 5.1 A 4-bus system

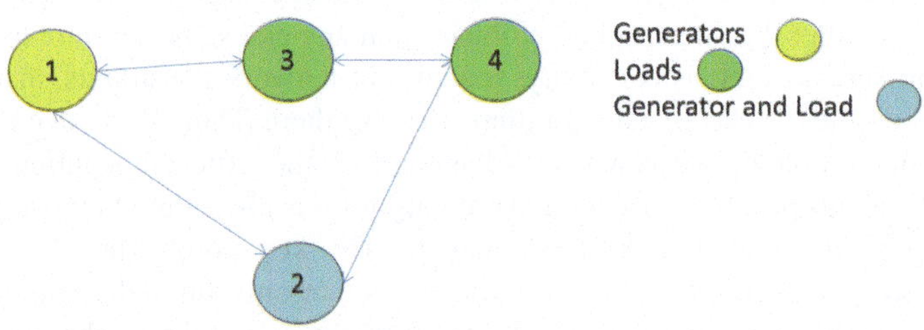

Fig. 5.2 A network model of the 4-bus system

constraint C6 points out that the total resistance of these buses should not exceed 17 MΩ.

Let the objective function be defined as follows:

$$\text{maximize power}: 6x_1 + 5x_2 + 3x_3 + 4x_4;$$

Subject to

$x_1 + x_2 \leq 5$ kV;	Constraint#1
$3x_1 + 2x_2 \leq 12$ h per year;	Constraint#2
$x_3 + 2x_4 \leq 8$ kV;	Constraint#3
$2x_3 + x_4 \leq 10$ h per year;	Constraint#4
$x_1 + x_2 + x_3 + x_4 \leq 7$ kw;	Constraint#5
$2x_1 + x_2 + x_3 + 3x_4 \leq 17\ M\Omega$;	Constraint#6

We can compute the border-feasible points (corner points) using AMPL software directly or using a graphical approach. The corner points for the first two equations (i.e., C1 and C2) are (0, 0), (0,5), (2,3), and (4,0), and of the four points, point (2,3) maximizes the objective function with a value of $Z = 27$. Similarly, the corner points for the next two equations (i.e., C3 and C4) are (0, 0) (5, 0), (0, 4), and (4, 2), and of the four points, point (4, 2) maximizes the objective function with a value of $Z = 20$. Constraints C5 and C6 are called master constraints.

$$\begin{aligned}
&\text{minimize power}: -6x_1 - 5x_2 - 3x_3 - 4x_4;\\
&\text{minimize}: \sum C_j \lambda_j X_j\\
&\text{subject to } \sum A\lambda_j X_j \leq b_i;\\
&\text{subject to } \sum \lambda_j = 1;\\
&\text{i.e.,} Y\,P_j - C_j > 0;
\end{aligned}$$

Considering dual variables (Y) and the entering column (P_j), we can formulate the stopping criteria for the DW algorithm to terminate as follows:

$$\begin{aligned}
&(w, \alpha)\begin{pmatrix} Ax_j \\ 1 \end{pmatrix} - cX_j > 0;\\
&(w, \alpha - C)X_j + \alpha > 0;
\end{aligned}$$

where y [i.e., w and α] is the dual variable, $p_j =$ entering column, $c_j =$ objective function coefficients.

Let us introduce two slack variables, S_1 and S_2, to the master problem constraints because these variables do not have any impact on the objective function or on the optimal values. We create the identity matrix shown in Table 5.1 for S_1, S_2, and λ_1.

Table 5.1 Simplex table for the Dantzig-Wolfe for a 4-bus system

Variables	S_1	S_2	λ_1	RHS	$\begin{pmatrix} Ax \\ 1 \end{pmatrix}$	θ	θ
S_1	1	0	0	7	11	7/11	
S_2	0	1	0	17	17	1	
λ_1	0	0	1	1	1	1	
$Z-C_j$	0	0	0	0	47		
λ_2	1/11	0	0	7/11	1	4/11	7/4
S_2	−17/11	1	0	68/11	0	20/11	17/5
λ_1	−1/11	0	1	4/11	0	7/11	4/7
$Z-C_j$	−47/11	0	0	−329/11	76/11		
λ_2	1/7	0	−4/7	3/7	0	4/7	3/4
S_2	−9/7	1	−20/7	36/7	0	48/7	3/4
λ_3	−1/7	0	11/7	4/7	1	3/7	4/3
$Z-C_j$	−23/7	0	−76/7	−237/7		20/7	
λ_4	¼	0	−1	¾	1		
S_2	−3	1	4	0	0		
λ_3	−1/4	0	2	¼	0		
	−4	0	8	36			

We then assign the right-hand side coefficients of the master constraints (C5 and C6) toward the RHS column. The coefficients of the master constraints are in the A matrix, and the initial corner values for x are taken as (2,3,4,2). The resultant product is (11, 17).

Table 5.1 shows the procedure to compute optimal values using the Dantzig-Wolfe technique. Table 5.1 has two slack variables, S_1 and S_2, and four convexity constraints are needed to attain an optimal solution for the above-mentioned constraints and objective function. The table provides analysis for attaining a basic, feasible solution and determining which variable leaves the basis. The computation step for the row and column operations is explained after the table.

$$\begin{pmatrix} Ax_j \\ 1 \end{pmatrix} = \begin{pmatrix} 1 & 1 & 1 & 1 \\ 2 & 1 & 1 & 3 \end{pmatrix} \begin{pmatrix} 2 \\ 3 \\ 4 \\ 2 \end{pmatrix} = \begin{pmatrix} 11 \\ 17 \\ 1 \end{pmatrix}$$

The minimum value of θ leaves the basis, hence row S_1 with $\theta = 7/11$ is replaced with λ_2 in the next set. Then, a set of row operations is performed in the following sequence.

$$\lambda_2 = \frac{s_1}{11}; S_2 = (s_2 - 17\lambda_2); \lambda_1(\text{new}) = \lambda_1 - \lambda_2; Z - C = Z_c - 47\lambda_2$$

Using the termination condition of the DW technique, we solve for

Max $(WA - C)x + \alpha$:

$$WA = (-47/11, 0)\begin{pmatrix} 1 & 1 & 1 & 1 \\ 2 & 1 & 1 & 3 \end{pmatrix}$$

$$WA = (\,-47/11 \quad -47/11 \quad -47/11 \quad -47/11\,)$$

$$WA - C = (\,19/11 \quad 8/11 \quad -14/11 \quad -3/11\,)$$

Hence, the current objective function is modified as follows:

$$\frac{19}{11}x_1 + \frac{8}{11}x_2 - \frac{14}{11}x_3 - \frac{3}{11}x_4$$

Let us apply the same corner points to the new objective function, (0,0), (0,5), (2,3) and (4,0), to get a point that maximizes the objective function. Here, it is (4,0) for the first two constraints. Similarly, (0,0) maximizes constraints c3 and c4 from border points (0,0), (5,0), (0,4), and (4,2).

Then, the current corner-point values are $X_j = (4{,}0{,}0{,}0)$ and maximize at $Z = 76/11$. Again, let us calculate the new entering column information, P_j, that will enable us to obtain $\begin{pmatrix} Ax_j \\ 1 \end{pmatrix}$.

$$P_j = \begin{pmatrix} Ax_j \\ 1 \end{pmatrix} = \begin{pmatrix} 1 & 1 & 1 & 1 \\ 2 & 1 & 1 & 3 \end{pmatrix}\begin{pmatrix} 4 \\ 0 \\ 0 \\ 0 \end{pmatrix} = \begin{pmatrix} 4 \\ 8 \\ 1 \end{pmatrix};$$

$$\mathrm{B}^{-1}\begin{pmatrix} Ax_j \\ 1 \end{pmatrix} = \begin{pmatrix} 1/11 & 0 & 0 \\ -17/11 & 1 & 0 \\ -1/11 & 0 & 1 \end{pmatrix}\begin{pmatrix} 4 \\ 8 \\ 1 \end{pmatrix} = \begin{pmatrix} 4/11 \\ 20/11 \\ 7/11 \end{pmatrix};$$

The value of dual variables w and α is noted (−47/11,0, 0) to improve the objective function. This process continues until an optimal solution is reached as shown in Table 5.2.

As noted in Table 5.2, corner point (4,0,0,0) repeats, and we have obtained the preferred objective function to have a value of 0. This task terminates the algorithm, enabling the calculation of the final corner point that maximizes our objective function, resulting in Z = 36. An AMPL output using a direct LP implementation for our 4-bus example is shown in Fig. 5.3.

Table 5.2 Final corner points and objective

x_1, x_2, x_3, x_4	Optimal value	Iteration
(0,0,0,0)	0	Start 0
(2,3,4,2)	47	Iteration 1
(4,0,0,0)	−76/11	Iteration 2
(4,0,0,4)	20/7	Iteration 3
(4,0,0,0)	0	Terminated

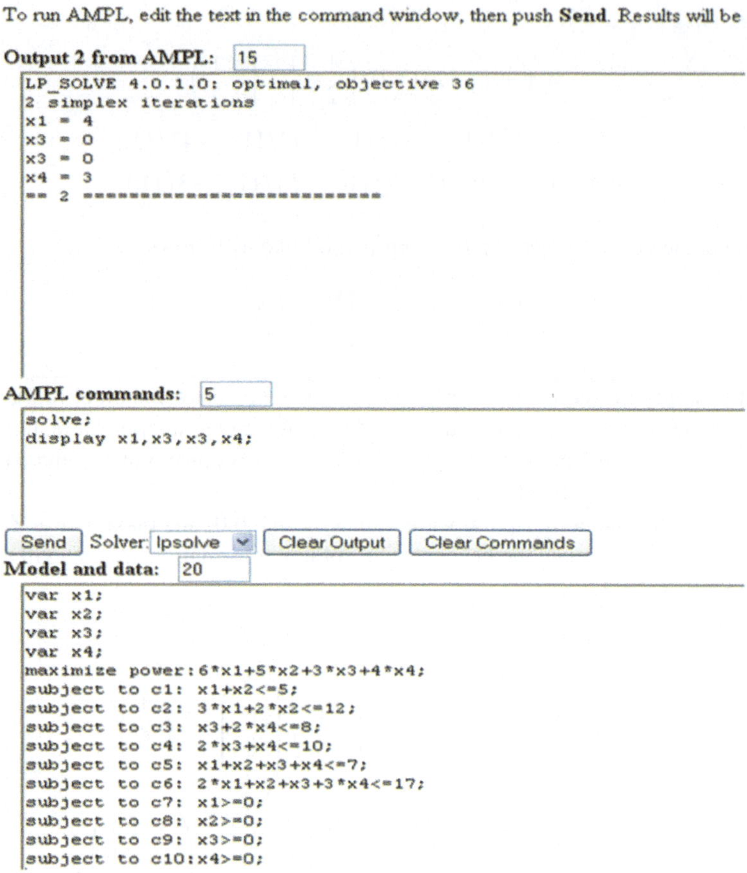

Fig. 5.3 An AMPL run of DLP

The data file has the following parameters: param x_1:=0; param x_2:=0; param x_3:=0; param x_4:=0.

Notice that the solution for both the direct LP and DW approaches is the same. The final corner values that buses can take to achieve their maximization objective can be calculated as

$= \lambda_3 {}^* x_3 + \lambda_4 {}^* x_4$.

= 1/4 * (4,0,0,0) + 3/4 * (4,0,0,4).

= (1,0,0,0) + (3,0,0,3).

Thus, x_1, x_2, x_3, x_4 = (4,0,0,3).

For these point combinations, the final optimal value results in Z = 36, hence Bus 1 and Bus 4 should be kept at 4 kv and 3 kv of generation to obtain the maximum power-flow capacity of 36 MW in the network. Thus, at each DW iteration, a relaxed version of the master problem is solved. Then, N sub-problems are solved using the reduced costs of the master linear program as parameters. As a

Table 5.3 Computational savings of the Dantzig-Wolfe method over the direct approach

Number of variables	Using LP directly (in seconds)	D-Wolfe (in seconds)	Constraints	Difference	Optimum power	Iterations using LP
4	0.000999	0.000183	10	8.16×10^{-4}	Z = 36	2
5	0.01999	0.000181	11	0.019802	Z = 39	3
6	0.000999	0.00017	12	0.01982	Z = 37.66	6
7	0.000999	0.00014	13	8.59×10^{-4}	Z = 37	7
8	0.000999	0.00014	14	8.59×10^{-4}	Z = 37.1	9

result, each sub-problem generates a candidate variable that is introduced in the master problem. The current relaxed master problem is updated by including all candidate variables that were found by the sub-problems. We code both the Dantzig-Wolfe technique and direct LP formulation for the same 4-bus problem and observe the computational savings as shown in Table 5.3.

Thus, Dantzig-Wolfe decomposition is an efficient optimization method when applied to large-scale problems with a special block-angular structure. Unlike the sub-gradient method, the Dantzig-Wolfe decomposition method properly defines new Lagrangian multipliers for subsequent sub-problems. The fast and monotonic convergence is a distinct feature of the Dantzig-Wolfe decomposition [Dan63, Chv83].

5.3 LP Formulation of the IEEE 14-BUS System

The IEEE bus system is a common practice that the academic and research community uses to test new models. The data are readily available to develop models and to perform analysis. We have used data from the system to develop our formulations. We have formulated an LP model of the standard IEEE 14-bus system using the AMPL package. The basic system is implemented and tested with bus failure and repair rates to study the impact of line voltages and the buses' dynamic behavior. The failure-rate and repair-rate data for the IEEE 14 bus are taken from [WG10, Wan01] as shown in Tables 5.5–5.7. The objective for our formulation is to minimize the failure rate and repair rate subject to flow-balance constraints and capacity constraints. For simplicity, we have multiplied the failure rate and repair rate, defining the objective function variable as a "risk" or "loss." Hence, the goal is to minimize the risk of any energy loss for the IEEE bus system that is subjected to populated constraints.

A single-line diagram for the IEEE 14-bus standard system extracted from [Son99, Moo91] is shown in Fig. 5.4. It consists of five generators with IEEE type-1 exciters, three of which are synchronous compensators that are only used for reactive power support. There are 11 loads in the system, totaling 258 MW and 81.3 Mvar. For our analysis, we have only taken the supply's real power. The

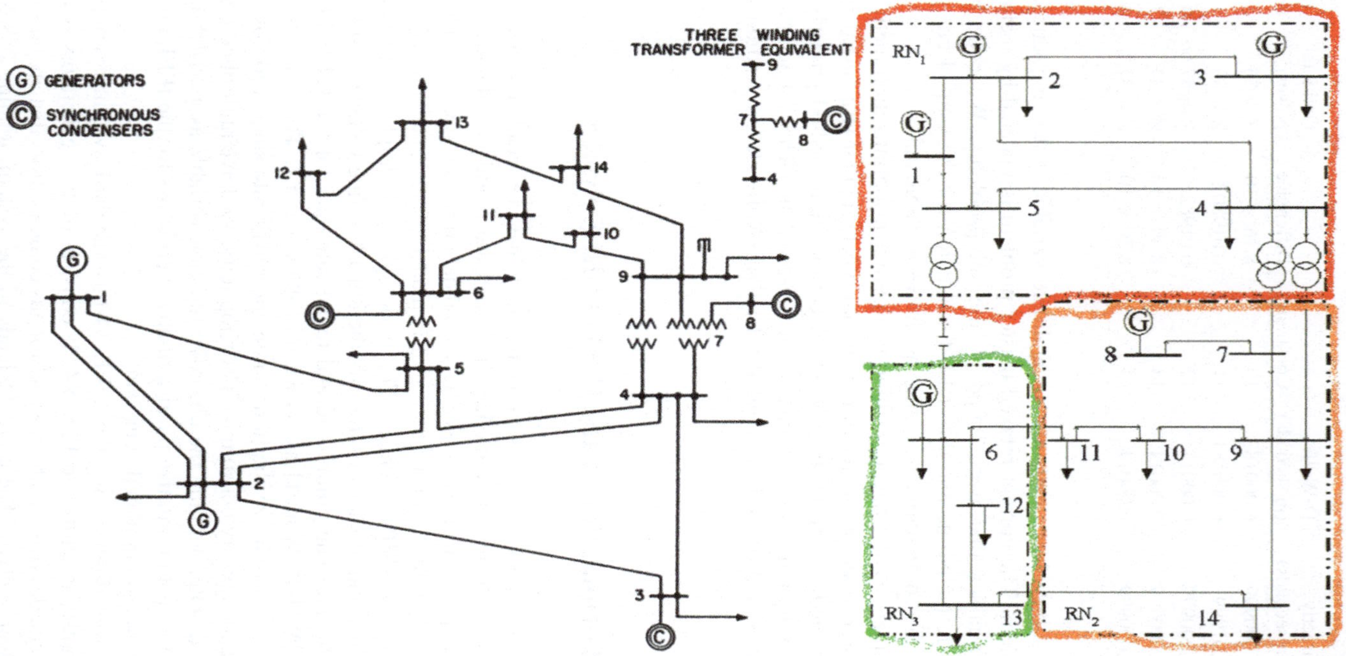

Fig. 5.4 IEEE 14-bus system and three decomposed regions: RN_1, RN_2, and RN_3

supply and demand in the IEEE 14-bus system equal 258 MW. Hence, it is a balanced system with equal values for the supply and demand units. Dynamic data for the generator exciters are selected from [WG10].

The IEEE bus system shown in Fig. 5.4 is decomposed into three regions, RN_1, RN_2, and RN_3, as shown in Fig. 5.5. The IEEE 14-bus network model contains 14 nodes and 18 transmission lines. A node is similar to a bus or point junction where two or multiple lines interconnect.

For example, lines 12 and 11 interconnect at node 6, and lines 14 and 12 connect at line 13. We have decomposed the system into three regions as done in a previous reliability study [WG10]. We may adapt random decomposition when choosing lines. In this system, we treat transmission lines 1, 2, 3, 4, and 5 as region 1; lines 7, 8, 9, 10, 11, and 14 as region 2; and lines 6, 12, and 13 as region 3.

The number of generators varies in each region of the IEEE 14-bus system. For example, there are three generators in region 1 (G1, G2, and G3), one generator in region 3 (G4), and one generator in region 2 (G5).

In real electric networks, transformers are used to interconnect multiple regions. For our analysis, we introduce new nodes, A1, A2, B1, B2, C1, and C2, that interconnect regions. This structure is shown in Fig. 5.6. For example, nodes A2 and A1 interconnect regions 1 and 3, and nodes C1 and C2 connect regions 1 and 2. Similarly, nodes B1 and B2 interconnect regions 3 and 1. We include these nodes and apply flow-balance constraints in our formulation.

In Fig. 5.6, the demand units (nodes) are represented as yellow circles, and the generators are not colored. The individual regions that interconnect with the key nodes are shown separately in Figs. 5.7–5.9, respectively. For example, only nodes A1, A2, C1, and C2 from other regions are involved in our analysis for the region-1 study. These nodes represent the total supply and demand allocation from their respective regions, hence it is not necessary to include all nodes.

As seen in Fig. 5.6, we introduce new artificial nodes, R1, R2, and R3, in the IEEE 14-bus system to keep excess power from reaching other regions and nodes. The presence of this node is due to the fact that we assume any loads can take power from the five generators in all three regions. By adding these new R nodes, we make sure that network follows the restriction on the available supply and demand values as per the IEEE 14-bus data. For example, joint capacities at node 4 should not exceed the demand of G41. The node G41 has an initial allocation of 16.5 MW, thus the contribution of generator 4 to region 1 should not exceed 16.5 MW. A negative sign in the R1 node for G41 indicates that it is a demand. Figure 5.7 shows the R1 node constraints for region 1.

We outline our basic formulation for Direct LP and Dantzig-Wolfe in Table 5.4. The term "Direct LP" refers to allocating the IEEE 14 bus into regions without any decomposition. The Dantzig-Wolfe formulation has variables d and f, indicating the sub-problems or regional constraints. In Direct LP, we combine all constraints in $Ax = b$ notation, whereas with DW, only the master constraints are constructed. We will discuss the details in the implementation section of Chap. 6.

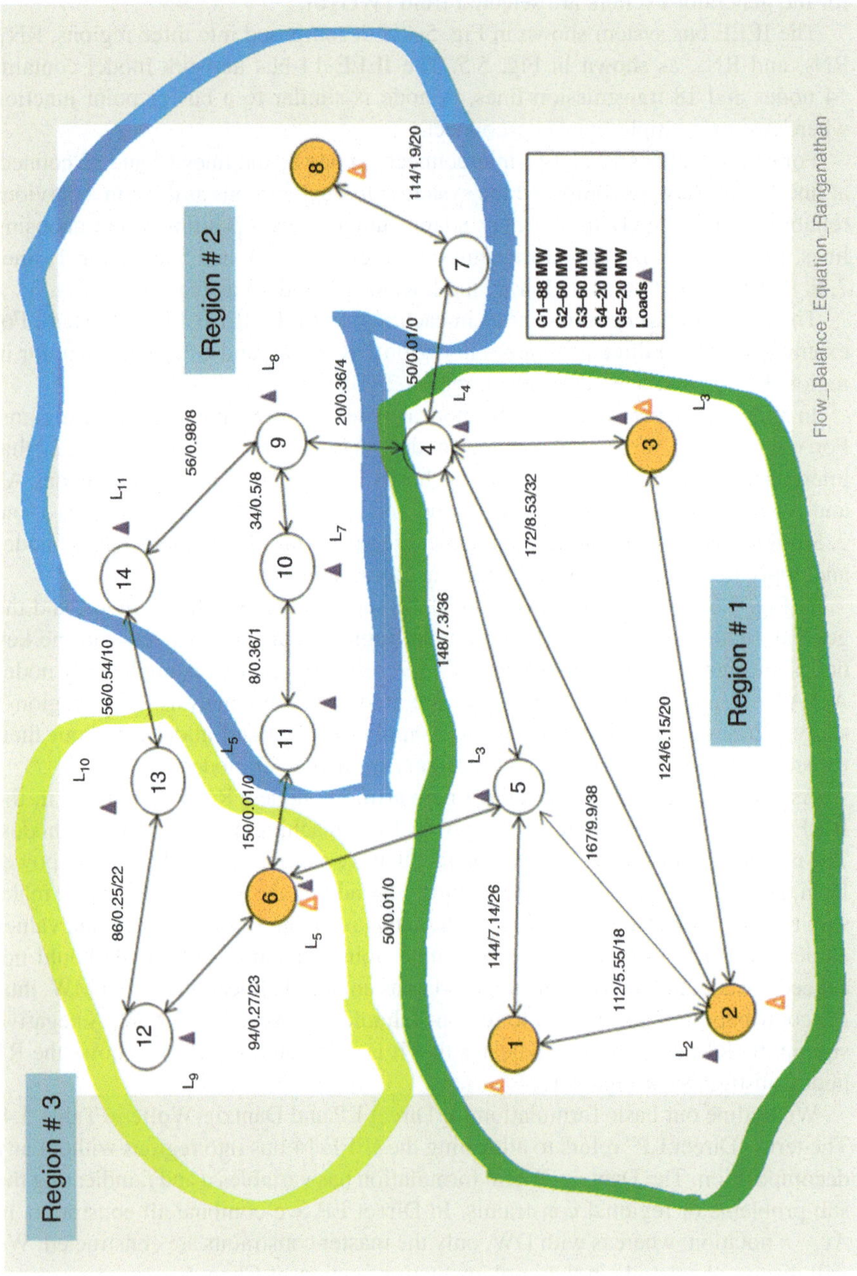

Fig. 5.5 Network model of Regional decomposition

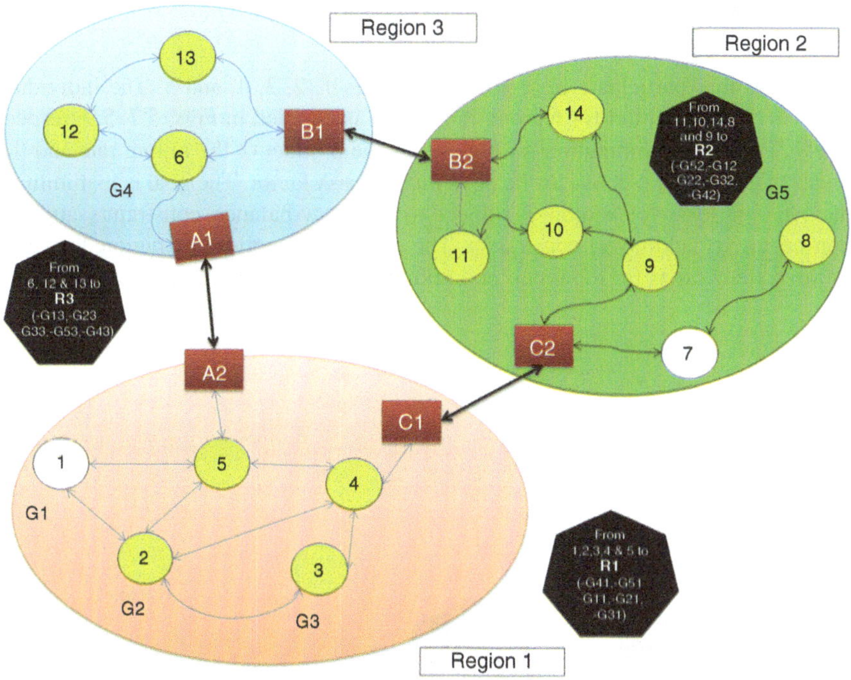

Fig. 5.6 IEEE 14-bus network model with local R constraints

Table 5.4 LP formulation of the Dantzig-Wolfe and direct approaches

Direct LP formulation	Dantzig-Wolfe decomposition formulation
min $Z = Cx$	min $Z = C\lambda x$
s.t. $Ax \leq b$; (Sub-problems 1,2…n and Master constraints)	s.t. $A\lambda x \leq b$; (Master constraint)
	s.t. $dx \leq f$; (Sub-problem 1,2,3…n)
s.t. $x \geq 0$;	s.t. $\sum_k \lambda_k = 1$; (convexity constraint)
	s.t. $x \geq 0$;
where x is the decision variable	where x is the decision variable

5.3.1 *Region 1 Constraints*

The nodes that participate in region 1 are nodes 1, 2, 3, 4, and 5. The individual regional decomposition for regions 1, 3, and 2 is detailed in Figs. 5.7–5.9, respectively. The objective function for region 1 is the product of the failure rate and the repair rate, which we define as the power loss or risk factor. The goal is to minimize the risk, or loss, for region 1 subject to the flow-balance constraints and the non-negativity additional constraints. The objective function for region 1 as well as the actual values for the repair and failure rates are given in Table 5.5.

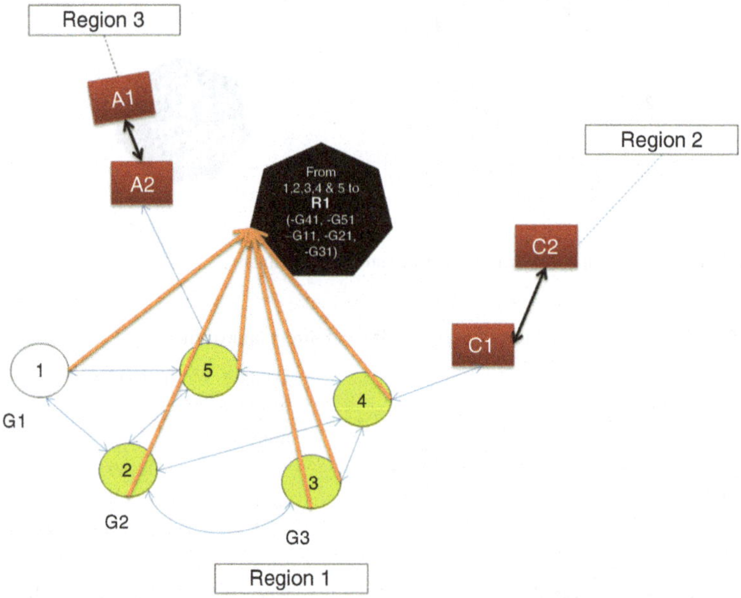

Fig. 5.7 R1 node constraint for nodes in Region 1

Table 5.5 Failure and Repair rates for Region 1

Lines connecting	Failure rate (λ)	Repair rate (r)
1–2	5.5552	18
1–5	7.1424	26
2–3	6.1504	20
2–4	9.9200	38
2–5	6.3488	22
3–4	8.5312	32
4–5	7.3408	36

5.3.1.1 Objective for Region 1 (Z_{LOSS})

$$Z_{\mathrm{LOSS}-R1} = 100x_{1-2-1} + 186x_{1-5-1} + 123x_{2-3-1} + 376x_{2-4-1} + 140x_{2-5-1} + 273x_{3-4-1} + 262x_{4-5-1} + 100x_{1-2-2} + 186x_{1-5-2} + 123x_{2-3-2} + 376x_{2-4-2} + 140x_{2-5-2} + 273x_{3-4-2} + 262x_{4-5-2} + 100x_{1-2-3} + 186x_{1-5-3} + 123x_{2-3-3} + 376x_{2-4-3} + 140x_{2-5-3} + 273x_{3-4-3} + 262x_{4-5-3} + 100x_{1-2-4} + 186x_{1-5-4} + 123x_{2-3-4} + 376x_{2-4-4} + 140x_{2-5-4} + 273x_{3-4-4} + 262x_{4-5-4} + 100x_{1-2-5} + 186x_{1-5-5} + 123x_{2-3-5} + 376x_{2-4-5} + 140x_{2-5-5} + 273x_{3-4-5} + 262x_{4-5-5};$$

The flow-balance constraints in region 1 are formulated at each node as follows:

5.3.1.2 Node 1

Here, there is no demand in node 1, so we assign zero on the right-hand side.

$$(x_{1-2-1} + x_{1-5-1}) - (x_{2-1-1} + x_{5-1-1}) + x_{1-R1-1} = 0; \tag{5.1}$$

$$(x_{1-2-2} + x_{1-5-2}) - (x_{2-1-2} + x_{5-1-2}) + x_{1-R1-2} = 0; \tag{5.2}$$

$$(x_{1-2-3} + x_{1-5-3}) - (x_{2-1-3} + x_{5-1-3}) + x_{1-R1-3} = 0; \tag{5.3}$$

$$(x_{1-2-4} + x_{1-5-4}) - (x_{2-1-4} + x_{5-1-4}) + x_{1-R1-4} = 0; \tag{5.4}$$

$$(x_{1-2-5} + x_{1-5-5}) - (x_{2-1-5} + x_{5-1-5}) + x_{1-R1-5} = 0; \tag{5.5}$$

5.3.1.3 Node 2

The flow-balance constraints for node 2 in region 1 are formulated at each node as follows:

$$(x_{2-1-1} + x_{2-5-1} + x_{2-4-1} + x_{2-3-1}) - (x_{1-2-1} + x_{5-2-1} + x_{4-2-1} + x_{3-2-1}) + x_{2-R1-1} \leq G_{11} - 21.7; \tag{5.6}$$

$$(x_{2-1-2} + x_{2-5-2} + x_{2-4-2} + x_{2-3-2}) - (x_{1-2-2} + x_{5-2-2} + x_{4-2-2} + x_{3-2-2}) + x_{2-R1-2} \leq G_{21} - 21.7; \tag{5.7}$$

$$(x_{2-1-3} + x_{2-5-3} + x_{2-4-3} + x_{2-3-3}) - (x_{1-2-3} + x_{5-2-3} + x_{4-2-3} + x_{3-2-3}) + x_{2-R1-3} \leq G_{31} - 21.7; \tag{5.8}$$

$$\begin{aligned}&(x_{2-1-4}+x_{2-5-4}+x_{2-4-4}+x_{2-3-4})\\&\quad-(x_{1-2-4}+x_{5-2-4}+x_{4-2-4}+x_{3-2-4})+x_{2-R1-4}\\&\quad\le G_{41}-21.7;\end{aligned} \tag{5.9}$$

$$\begin{aligned}&(x_{2-1-5}+x_{2-5-5}+x_{2-4-5}+x_{2-3-5})\\&\quad-(x_{1-2-5}+x_{5-2-5}+x_{4-2-5}+x_{3-2-5})+x_{2-R1-5}\\&\quad\le G_{51}-21.7;\end{aligned} \tag{5.10}$$

5.3.1.4 Node 3

The flow-balance constraints at node 3 in region 1 are formulated at each node as follows:

$$(x_{3-2-1}+x_{3-4-1})-(x_{2-3-1}+x_{4-3-1})+x_{3-R1-1}\le G_{11}-94.2; \tag{5.11}$$

$$(x_{3-2-2}+x_{3-4-2})-(x_{2-3-2}+x_{4-3-2})+x_{3-R1-2}\le G_{21}-94.2; \tag{5.12}$$

$$(x_{3-2-3}+x_{3-4-3})-(x_{2-3-3}+x_{4-3-3})+x_{3-R1-3}\le G_{31}-94.2; \tag{5.13}$$

$$(x_{3-2-4}+x_{3-4-4})-(x_{2-3-4}+x_{4-3-4})+x_{3-R1-4}\le G_{41}-94.2; \tag{5.14}$$

$$(x_{3-2-5}+x_{3-4-5})-(x_{2-3-5}+x_{4-3-5})+x_{3-R1-5}\le G_{51}-94.2; \tag{5.15}$$

5.3.1.5 Node 4

The flow-balance constraints at node 4 in region 1 are formulated at each node as follows:

$$\begin{aligned}&(x_{4-5-1}+x_{4-2-1}+x_{4-3-1})-(x_{5-4-1}+x_{2-4-1}+x_{3-4-1})+x_{4-R1-1}\\&\quad\le G_{11}-47;\end{aligned} \tag{5.16}$$

$$\begin{aligned}&(x_{4-5-2}+x_{4-2-2}+x_{4-3-2})-(x_{5-4-2}+x_{2-4-2}+x_{3-4-2})+x_{4-R1-2}\\&\quad\le G_{21}-47;\end{aligned} \tag{5.17}$$

$$\begin{aligned}&(x_{4-5-3}+x_{4-2-3}+x_{4-3-3})-(x_{5-4-3}+x_{2-4-3}+x_{3-4-3})+x_{4-R1-3}\\&\quad\le G_{31}-47;\end{aligned} \tag{5.18}$$

$$\begin{aligned}&(x_{4-5-4}+x_{4-2-4}+x_{4-3-4})-(x_{5-4-4}+x_{2-4-4}+x_{3-4-4})+x_{4-R1-4}\\&\quad\le G_{41}-47;\end{aligned} \tag{5.19}$$

$$\begin{aligned}&(x_{4-5-5}+x_{4-2-5}+x_{4-3-5})-(x_{5-4-5}+x_{2-4-5}+x_{3-4-5})+x_{4-R1-5}\\&\quad\le G_{51}-47;\end{aligned} \tag{5.20}$$

5.3.1.6 Node 5

The flow-balance constraints at node 5 in region 1 are formulated at each node as follows:

$$\begin{aligned}&(x_{5-1-1}+x_{5-2-1}+x_{5-4-1}+x_{5,a2,1})\\&\quad-(x_{1-5-1}+x_{2-5-1}+x_{4-5-1}+x_{a2,5,4})+x_{5-R1-1}\\&\quad\leq G_{11}-7.6;\end{aligned} \tag{5.21}$$

$$\begin{aligned}&(x_{5-1-2}+x_{5-2-2}+x_{5-4-2}+x_{5,a2,2})\\&\quad-(x_{1-5-2}+x_{2-5-2}+x_{4-5-2}+x_{a2,5,4})+x_{5-R1-2}\\&\quad\leq G_{21}-7.6;\end{aligned} \tag{5.22}$$

$$\begin{aligned}&(x_{5-1-3}+x_{5-2-3}+x_{5-4-3}+x_{5,a2,3})\\&\quad-(x_{1-5-3}+x_{2-5-3}+x_{4-5-3}+x_{a2,5,4})+x_{5-R1-3}\\&\quad\leq G_{31}-7.6;\end{aligned} \tag{5.23}$$

$$\begin{aligned}&(x_{5-1-4}+x_{5-2-4}+x_{5-4-4}+x_{5,a2,4})\\&\quad-(x_{1-5-4}+x_{2-5-4}+x_{4-5-4}+x_{a2,5,4})+x_{5-R1-4}\\&\quad\leq G_{41}-7.6;\end{aligned} \tag{5.24}$$

$$\begin{aligned}&(x_{5-1-5}+x_{5-2-5}+x_{5-4-5}+x_{5,a2,5})\\&\quad-(x_{1-5-5}+x_{2-5-5}+x_{4-5-5}+x_{a2,5,4})+x_{5-R1-5}\\&\quad\leq G_{51}-7.6;\end{aligned} \tag{5.25}$$

5.3.1.7 Joint-Capacity Constraints for Region 3

The joint-capacity constraints in region 1 are formulated at each node as follows:

$$x_{1-R1-1}+x_{2-R1-1}+x_{3-R1-1}+x_{4-R1-1}+x_{5-R1-1}\leq -G_{11}; \tag{5.26}$$

$$x_{1-R1-2}+x_{2-R1-2}+x_{3-R1-2}+x_{4-R1-2}+x_{5-R1-2}\leq -G_{21}; \tag{5.27}$$

$$x_{1-R1-3}+x_{2-R1-3}+x_{3-R1-3}+x_{4-R1-3}+x_{5-R1-3}\leq -G_{31}; \tag{5.28}$$

$$x_{1-R1-4}+x_{2-R1-4}+x_{3-R1-4}+x_{4-R1-4}+x_{5-R1-4}\leq -G_{41}; \tag{5.29}$$

$$x_{1-R1-5}+x_{2-R1-5}+x_{3-R1-5}+x_{4-R1-5}+x_{5-R1-5}\leq -G_{51}; \tag{5.30}$$

5.3.1.8 Other Constraints

The other constraints in region 1 are formulated at each node as follows:

At A1,

$$x_{6,a1,4} - (x_{a1,a2,4}) = 0; \tag{5.31}$$

$$x_{a2,a1,1} - (x_{a1,6,1}) = 0; \tag{5.32}$$

$$x_{a2,a1,2} - (x_{a1,6,2}) = 0; \tag{5.33}$$

At A2,

$$(x_{a2,a1,1}) = 0;$$

$$x_{5,a2,2} - x_{a2,a1,2} = 0; \tag{5.34}$$

$$x_{5,a2,3} - x_{a2,a1,3} = 0 \tag{5.35}$$

$$x_{a1,a2,4} - (x_{a2,5,4}) = 0; \tag{5.36}$$

$$x_{5,\ a2,3} = G_{33}; \tag{5.37}$$

$$x_{5,a2,2} = G_{23}; \tag{5.38}$$

$$x_{5,a2,1} = G_{13}; \tag{5.39}$$

$$x_{5,a2,4} = G_{43}; \tag{5.40}$$

$$x_{5,a2,5} = G_{53}; \tag{5.41}$$

$$x_{a2,5,4} = G_{41}; \tag{5.42}$$

[END OF SUB PROBLEM 1].

5.3.2 *Region 3 Constraints (Nodes 6, 12, and 13)*

The nodes that participate in region 3 are nodes 6, 12, and 13. The individual regional decomposition for region 3 is detailed in Fig. 5.8. The objective function for region 3 is the product of the failure rate and the repair rate, which we define as the power loss or risk factor. The goal is to minimize the risk, or loss, for region 3 subject to the flow-balance constraints and non-negativity additional constraints. The following equation is the objective function for region 3, and the actual values for the repair and failure rates are given in Table 5.5.

5.3.2.1 Objective for Region 3 (Z_{LOSS})

The objective function for Region 3 can be written as

$$\begin{aligned} Z_{LOSS-R3} = {} & 6.3x_{6-12-1} + 5.5x_{12-13-1} + 6.3x_{6-12-2} + 5.5x_{12-13-2} + 6.3x_{6-12-3} \\ & + 5.5x_{12-13-3} + 6.3x_{6-12-4} + 5.5x_{12-13-4} + 6.3x_{6-12-5} \\ & + 5.5x_{12-13-5}; \end{aligned}$$

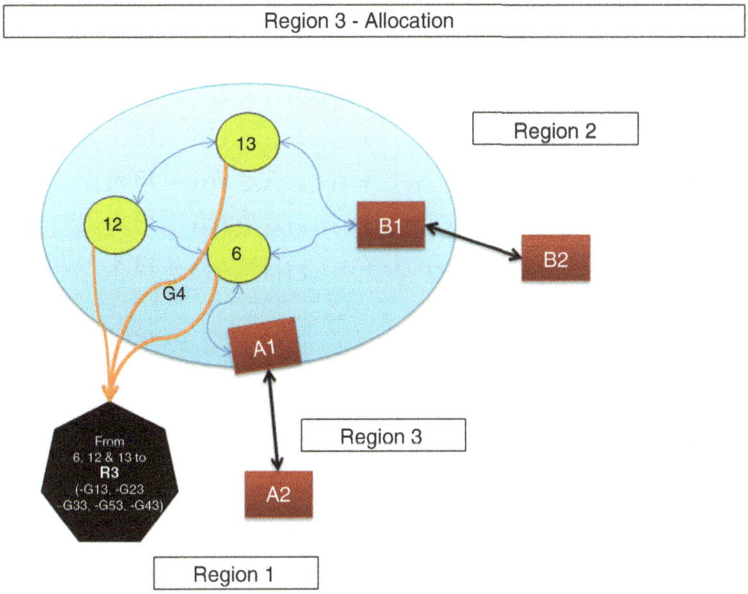

Fig. 5.8 R3 node constraints for nodes in Region 3

Table 5.6 Failure and repair rates for Region 3

Lines connecting	Failure rate (λ)	Repair rate (r)
6–12	0.274	23
6–13	0.44	0.44
12–13	0.25	22

The data to calculate the coefficients for objective function are taken from Table 5.6.

The flow-balance constraints in region 3 are formulated at each node and given as follows:

5.3.2.2 Node 12

$$x_{12-13-1} + x_{12-6-1} - (x_{13-12-1} + x_{6-12-1}) + x_{12\sim R3-1}) \leq G_{13} - 6; \qquad (5.43)$$

$$x_{12-13-2} + x_{12-6-2} - (x_{13-12-2} + x_{6-12-2}) + x_{12\sim R3-2}) \leq G_{23} - 6; \qquad (5.44)$$

$$x_{12-13-3} + x_{12-6-3} - (x_{13-12-3} + x_{6-12-3}) + x_{12\sim R3-3}) \leq G_{33} - 6; \qquad (5.45)$$

$$x_{12-13-4} + x_{12-6-4} - (x_{13-12-4} + x_{6-12-4}) + x_{12\sim R3-4}) \leq G_{43} - 6; \qquad (5.46)$$

$$x_{12-13-5} + x_{12-6-5} - (x_{13-12-5} + x_{6-12-5}) + x_{12\sim R3-5}) \leq G_{53} - 6; \qquad (5.47)$$

5.3.2.3 Node 13

$$x_{13-12-1} + x_{13,b1,4} - (x_{12-13-1}) + x_{13-R3-1} \leq G_{13} - 13.5; \quad (5.48)$$
$$x_{13-12-2} + x_{13,b1,4} - (x_{12-13-2}) + x_{13-R3-2} \leq G_{23} - 13.5; \quad (5.49)$$
$$x_{13-12-3} + x_{13,b1,4} - (x_{12-13-3}) + x_{13-R3-3} \leq G_{33} - 13.5; \quad (5.50)$$
$$x_{13-12-4} + x_{13,b1,4} - (x_{12-13-4}) + x_{13-R3-4} \leq G_{43} - 13.5; \quad (5.51)$$
$$x_{13-12-5} + x_{13,b1,4} - (x_{12-13-5}) + x_{13-R3-5} \leq G_{53} - 13.5; \quad (5.52)$$

5.3.2.4 Node 6

$$(x_{6-12-1}) - (x_{12-6-1}) + x_{6-R3-1} \leq G_{13} - 11.2; \quad (5.53)$$
$$(x_{6-12-2}) - (x_{12-6-2}) + x_{6-R3-2} \leq G_{23} - 11.2; \quad (5.54)$$
$$(x_{6-12-3}) - (x_{12-6-3}) + x_{6-R3-3} \leq G_{33} - 11.2; \quad (5.55)$$
$$(x_{6-12-4}) - (x_{12-6-4}) + x_{6-R3-4} \leq G_{43} - 11.2; \quad (5.56)$$
$$(x_{6-12-5}) - (x_{12-6-5}) + x_{6-R3-5} \leq G_{53} - 11.2; \quad (5.57)$$

5.3.2.5 Joint-Capacity Constraints for Region 3

$$x_{12-R3-1} + x_{13-R3-1} + x_{6-R3-1} \leq -G_{13}; \quad (5.58)$$
$$x_{12-R3-2} + x_{13-R3-2} + x_{6-R3-2} \leq -G_{23}; \quad (5.59)$$
$$x_{12-R3-3} + x_{13-R3-3} + x_{6-R3-3} \leq -G_{33}; \quad (5.60)$$
$$x_{12-R3-4} + x_{13-R3-4} + x_{6-R3-4} \leq -G_{43}; \quad (5.61)$$
$$x_{12-R3-5} + x_{13-R3-5} + x_{6-R3-5} \leq -G_{53}; \quad (5.62)$$

5.3.2.6 Other Constraints

$$x_{6,b1,4} - (x_{b1,6,4}) = 0; \quad (5.63)$$
$$x_{6,a1,4} - x_{a1,6,4} = 0; \quad (5.64)$$

A1:

$$x_{6,a1,4} - (x_{a1,a2,4}) = 0; \tag{5.65}$$

$$x_{a2,a1,1} - (x_{a1,6,1}) = 0; \tag{5.66}$$

$$x_{a2,a1,2} - (x_{a1,6,2}) = 0; \tag{5.67}$$

A2:

$$x_{5,a2,1} - (x_{a2,a1,1}) = 0; x_{5,a2,3} - x_{a2,a1,3} = 0; \tag{5.68}$$

$$x_{a1,a2,4} - (x_{a2,5,4}) = 0; \tag{5.69}$$

$$x_{5,a2,2} - x_{a2,a1,2} = 0 \tag{5.70}$$

B1:

$$x_{13,b1,4} + x_{6,b1,4} - (x_{b1,b2,4}) = 0; \tag{5.71}$$

$$x_{b2,b1,5} - (x_{b1,6,5} + x_{b1,13,5}) = 0; \tag{5.72}$$

B2:

$$x_{14-b2-5} + x_{11-b2-5} - (x_{b2,b1,5}) = 0; \tag{5.73}$$

$$x_{b1,b2,4} - (x_{b2-14-4} + x_{b2-11-4}) = 0; \tag{5.74}$$

$$x_{6,b1,4} + x_{13,b1,4} = G_{42}; \tag{5.75}$$

$$x_{b1,b2,4} = G_{42}; \tag{5.76}$$

$$x_{b2,14,4} + x_{b2,11,4} = G_{42}; \tag{5.77}$$

$$x_{b2,b1,5} = x_{14-b2-5} + x_{11-b2-5};$$

[END OF SUB PROBLEM 2].

5.3.3 *Region 2 Constraints*

The nodes that participate in region 2 are nodes 7, 8, 9, 10, 11, and 14. The individual regional decomposition for region 2 is detailed in Fig. 5.9. The objective function for region 2 is the product of the failure rate and the repair rate, which we define as the power loss or risk factor. The goal is to minimize the risk, or loss, for region 2 subject to the flow-balance constraints and non-negativity additional constraints. The following equation is the objective function for region 2, and the actual values for the repair and failure rates are given in Table 5.7.

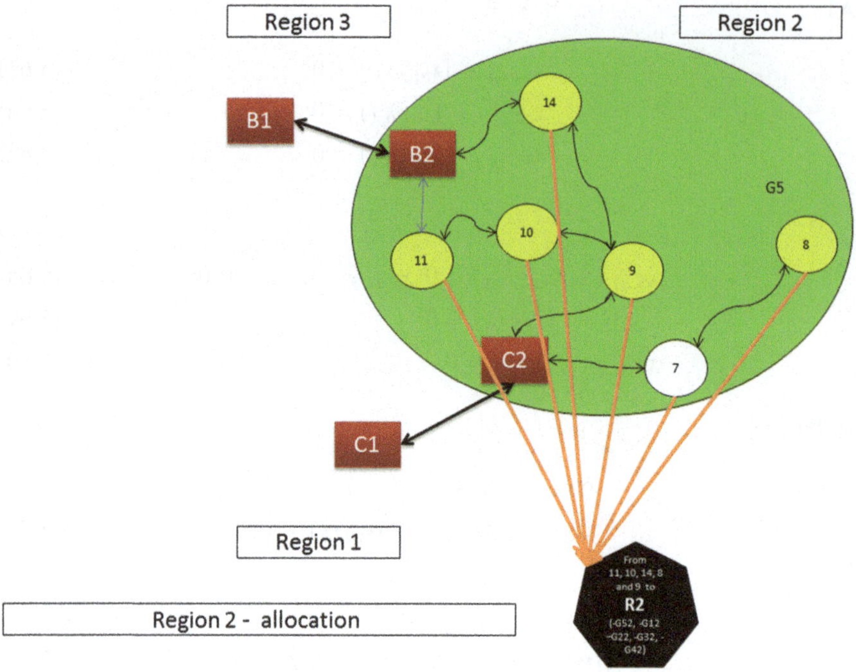

Fig. 5.9 R2 node constraints for nodes in Region 2

Table 5.7 Failure and repair rates for Region 2

Lines	Failure rate (λ)	Repair rate (r)
7–8	1.9973	20
7–10	0.946	15
7–11	1.08	13
8–9	1.506	14
8–10	1.121	12
9–10	0.595	8
9–14	0.981	8
10–11	0.38	6
10–14	1.29	16
10–0	1.12	13
14–0	0.85	11

5.3.3.1 Objective for Region 2 (Z_{LOSS})

The objective function for region 2 is given as

$$\begin{aligned} Z_{\mathrm{LOSS}-R2} = {} & 38x_{7-8-1} + 4x_{9-10-1} + 7.84x_{9-14-1} + 2.28x_{10-11-1} + 38x_{7-8-2} \\ & + 4x_{9-10-2} + 7.84x_{9-14-2} + 2.28x_{10-11-2} + 38x_{7-8-3} + 4x_{9-10-3} \\ & + 7.84x_{9-14-3} + 2.28x_{10-11-3} + 38x_{7-8-4} + 4x_{9-10-4} + 7.84x_{9-14-4} \\ & + 2.28x_{10-11-4} + 38x_{7-8-5} + 4x_{9-10-5} + 7.84x_{9-14-5} + 2.28x_{10-11-5}; \end{aligned}$$

5.3.3.2 Node 7

There is no demand at node 7.

$$(x_{C2,7,1} + x_{C2,7,2} + x_{C2,7,3}) - (x_{7,C2,5}) + x_{7-R2} = 0; \tag{5.78}$$

$$x_{8-7-1} - (x_{7-8-1}) = 0; \tag{5.79}$$

$$x_{8-7-2} - (x_{7-8-2}) = 0; \tag{5.80}$$

$$x_{8-7-3} - (x_{7-8-3}) = 0; \tag{5.81}$$

$$x_{8-7-4} - (x_{7-8-4}) = 0; \tag{5.82}$$

$$x_{8-7-5} - (x_{7-8-5}) = 0; \tag{5.83}$$

5.3.3.3 Node 8

There is no demand at node 8.

$$x_{8-7-1} - x_{7-8-1} + x_{8-R2-1} = 0; \tag{5.84}$$

$$x_{8-7-2} - x_{7-8-2} + x_{8-R2-2} = 0; \tag{5.85}$$

$$x_{8-7-3} - x_{7-8-3} + x_{8-R2-3} = 0; \tag{5.86}$$

$$x_{8-7-4} - x_{7-8-4} + x_{8-R2-4} = 0; \tag{5.87}$$

$$x_{8-7-5} - x_{7-8-5} + x_{8-R2-5} = 0; \tag{5.88}$$

5.3.3.4 Node 9

$$(x_{10-9-1} + x_{14-9-1}) - (x_{9-10-1} + x_{9-14-1}) + x_{9,c2,1} + x_{9-R2-1} \leq G_{12} - 29.5; \quad (5.89)$$

$$(x_{10-9-2} + x_{14-9-2}) - (x_{9-10-2} + x_{9-14-2}) + x_{9,c2,2} + x_{9-R2-2} \leq G_{22} - 29.5; \quad (5.90)$$

$$(x_{10-9-3} + x_{14-9-3}) - (x_{9-10-3} + x_{9-14-3}) + x_{9,c2,3} + x_{9-R2-3} \leq G_{32} - 29.5; \quad (5.91)$$

$$(x_{10-9-4} + x_{14-9-4}) - (x_{9-10-4} + x_{9-14-4}) + x_{9,c2,4} + x_{9-R2-4} \leq G_{42} - 29.5; \quad (5.92)$$

$$(x_{10-9-5} + x_{14-9-5}) - (x_{9-10-5} + x_{9-14-5}) + x_{9,c2,5} + x_{9-R2-5} \leq G_{52} - 29.5; \quad (5.93)$$

5.3.3.5 Node 10

$$x_{10-11-1} + x_{10-9-1} - (x_{11-10-1} + x_{9-10-1}) + x_{10-R2-1} \leq G_{12} - 9; \quad (5.94)$$

$$x_{10-11-2} + x_{10-9-2} - (x_{11-10-2} + x_{9-10-2}) + x_{10-R2-2} \leq G_{22} - 9; \quad (5.95)$$

$$x_{10-11-3} + x_{10-9-3} - (x_{11-10-3} + x_{9-10-3}) + x_{10-R2-3} \leq G_{32} - 9; \quad (5.96)$$

$$x_{10-11-4} + x_{10-9-4} - (x_{11-10-4} + x_{9-10-4}) + x_{10-R2-4} \leq G_{42} - 9; \quad (5.97)$$

$$x_{10-11-5} + x_{10-9-5} - (x_{11-10-5} + x_{9-10-5}) + x_{10-R2-5} \leq G_{52} - 9; \quad (5.98)$$

5.3.3.6 Node 11

$$x_{11-10-1} + x_{11,b2,1} - (x_{10-11-1} + x_{b2,11,1}) + x_{11-R2-1} \leq G_{12} - 3.5; \quad (5.99)$$

$$x_{11-10-2} + x_{11,b2,2} - (x_{10-11-2} + x_{b2,11,2}) + x_{11-R2-2} \leq G_{22} - 3.5; \quad (5.100)$$

$$x_{11-10-3} + x_{11,b2,3} - (x_{10-11-3} + x_{b2,11,3}) + x_{11-R2-3} \leq G_{32} - 3.5; \quad (5.101)$$

$$x_{11-10-4} + x_{11,b2,4} - (x_{10-11-4} + x_{b2,11,4}) + x_{11-R2-4} \leq G_{42} - 3.5; \quad (5.102)$$

$$x_{11-10-5} + x_{11,b2,5} - (x_{10-11-5} + x_{b2,11,5}) + x_{11-R2-5} \leq G_{52} - 3.5; \quad (5.103)$$

5.3.3.7 Node 14

$$x_{14-9-1} + x_{14,b2,1} - \left(x_{b2,14,1} + x_{9-14-1}\right) + x_{14-R2-1} \leq G_{12} - 14.8; \quad (5.104)$$
$$x_{14-9-2} + x_{14,b2,2} - \left(x_{b2,14,2} + x_{9-14-2}\right) + x_{14-R2-2} \leq G_{22} - 14.8; \quad (5.105)$$
$$x_{14-9-3} + x_{14,b2,3} - \left(x_{b2,14,3} + x_{9-14-3}\right) + x_{14-R2-3} \leq G_{32} - 14.8; \quad (5.106)$$
$$x_{14-9-4} + x_{14,b2,4} - \left(x_{b2,14,4} + x_{9-14-4}\right) + x_{14-R2-4} \leq G_{42} - 14.8; \quad (5.107)$$
$$x_{14-9-5} + x_{14,b2,5} - \left(x_{b2,14,5} + x_{9-14-5}\right) + x_{14-R2-5} \leq G_{52} - 14.8; \quad (5.108)$$

5.3.3.8 Joint-Capacity Constraints for Region 2

$$x_{14-R2-1} + x_{11-R2-1} + x_{10-R2-1} + x_{9-R2-1} + x_{7-R2-1} + x_{8-R2-1} \leq -G_{12}; \quad (5.109)$$
$$x_{14-R2-2} + x_{11-R2-2} + x_{10-R2-2} + x_{9-R2-2} + x_{7-R2-2} + x_{8-R2-2} \leq -G_{22}; \quad (5.110)$$
$$x_{14-R2-3} + x_{11-R2-3} + x_{10-R2-3} + x_{9-R2-3} + x_{7-R2-3} + x_{8-R2-3} \leq -G_{32}; \quad (5.111)$$
$$x_{14-R2-4} + x_{11-R2-4} + x_{10-R2-4} + x_{9-R2-4} + x_{7-R2-4} + x_{8-R2-4} \leq -G_{42}; \quad (5.112)$$
$$x_{14-R2-5} + x_{11-R2-5} + x_{10-R2-5} + x_{9-R2-5} + x_{7-R2-5} + x_{8-R2-5} \leq -G_{52}; \quad (5.113)$$

5.3.3.9 Other Constraints

B1:

$$x_{13,b1,4} + x_{6,b1,4} - \left(x_{b1,b2,4}\right) = 0; x_{b2,b1,5} - \left(x_{b1,6,5} + x_{b1,13,5}\right) = 0; \quad (5.114)$$

B2:

$$x_{14-b2-5} + x_{11-b2-5} - \left(x_{b2,b1,5}\right) = 0; \quad (5.115)$$
$$x_{b1,b2,4} - \left(x_{b2-14-4} + x_{b2-11-4}\right) = 0; \quad (5.116)$$

C1:

$$x_{4,c1,1} - (x_{c1,c2,1}) = 0; \quad (5.117)$$
$$x_{4,c1,2} - (x_{c1,c2,2}) = 0; \quad (5.118)$$
$$x_{4,c1,3} - (x_{c1,c2,3}) = 0; \quad (5.119)$$
$$x_{c2,c1,5} - (x_{c1,4,1} + x_{c1,4,2} + x_{c1,4,3}) = 0; \quad (5.120)$$

C2:

$$x_{9,c2,4} + x_{7,c2,4} - (x_{c2,c1,4}) = 0; \quad (5.121)$$
$$x_{9,c2,5} + x_{7,c2,5} - (x_{c2,c1,5}) = 0; \quad (5.122)$$
$$x_{c1,c2,1} - (x_{c2,9,1} + x_{c2,7,1}) = 0; \quad (5.123)$$
$$x_{c1,c2,2} - (x_{c2,9,2} + x_{c2,7,2}) = 0; \quad (5.124)$$
$$x_{c1,c2,3} - (x_{c2,9,3} + x_{c2,7,3}) = 0; \quad (5.125)$$
$$x_{a2,a1,3,} = G_{33}; \quad (5.126)$$
$$x_{a2,a1,2} = G_{23}; \quad (5.127)$$
$$x_{a2,a1,1=} G_{13}; \quad (5.128)$$
$$x_{a1,6,1} = G_{13}; \quad (5.129)$$
$$x_{a1,6,2} = G_{23}; \quad (5.130)$$
$$x_{a1,6,3} = G_{33}; \quad (5.131)$$

[END OF SUB PROBLEM 3].

5.3.4 Master Constraints (Linking Constraints)

The master constraints are the linking constraints that connect to sub-problems or regional constraints. The generator variables present in the master constraints are also included or related to the region 1, region 2, and region 3 constraints, as discussed above, during flow-balance and joint-capacity constraints. These constraints interact iteratively with the sub-constraints to reach an optimal solution via dual values and convexity constraints as proposed in the objective formulation. Constraints 153 to 158 serve as master constraints.

$$G_{11} + G_{12} + G_{13} = 88 \quad \text{(MC\#1)} \quad (5.132)$$
$$G_{21} + G_{22} + G_{23} = 60; \quad \text{(MC\#2)} \quad (5.133)$$
$$G_{31} + G_{32} + G_{33} = 60 \quad \text{(MC\#3)} \quad (5.134)$$

$$G_{41} + G_{42} + G_{43} = 25 \quad (\text{MC\#4}) \tag{5.135}$$

$$G_{51} + G_{52} + G_{53} = 25 \quad (\text{MC\#5}) \tag{5.136}$$

This five set of constraints means Commodity 1 Constraints <=88; Commodity 2 Constraints <=60; Commodity 3 Constraints <=60; Commodity 4 Constraints <=25; Commodity 5 Constraints <=25;

5.4 Decomposing the IEEE 14-Bus System into Two Regions

The constraints that change when modifying the three-region classifications into two regions are given in the following sections.

5.4.1 R2 Node Constraint in Region 1

$$\begin{aligned} &x_{12-R2-1} + x_{6-R2-1} + x_{13-R2-1} + x_{11-R2-1} + x_{10-R2-1} + x_{14-R2-1} \\ &\quad + x_{9-R2-1} + x_{8-R2-1} + x_{7-R2-1} \leq -G_{12}; \end{aligned} \tag{5.137}$$

$$\begin{aligned} &x_{12-R2-2} + x_{6-R2-2} + x_{13-R2-2} + x_{11-R2-2} + x_{10-R2-2} + x_{14-R2-2} \\ &\quad + x_{9-R2-2} + x_{8-R2-2} + x_{7-R2-2} \leq -G_{22}; \end{aligned} \tag{5.138}$$

$$\begin{aligned} &x_{12-R2-3} + x_{6-R2-3} + x_{13-R2-3} + x_{11-R2-3} + x_{10-R2-3} + x_{14-R2-3} \\ &\quad + x_{9-R2-3} + x_{8-R2-3} + x_{7-R2-3} \leq -G_{32}; \end{aligned} \tag{5.139}$$

$$\begin{aligned} &x_{12-R2-4} + x_{6-R2-4} + x_{13-R2-4} + x_{11-R2-4} + x_{10-R2-4} + x_{14-R2-4} \\ &\quad + x_{9-R2-4} + x_{8-R2-4} + x_{7-R2-4} \leq -G_{42}; \end{aligned} \tag{5.140}$$

$$\begin{aligned} &x_{12-R2-5} + x_{6-R2-5} + x_{13-R2-5} + x_{11-R2-5} + x_{10-R2-5} + x_{14-R2-5} \\ &\quad + x_{9-R2-5} + x_{8-R2-5} + x_{7-R2-5} \leq -G_{52}; \end{aligned} \tag{5.141}$$

5.4.2 R1 Node Constraint in Region 1

$$x_{1-R1-1} + x_{2-R1-1} + x_{3-R1-1} + x_{4-R1-1} + x_{5-R1-1} \leq -G_{11}; \tag{5.142}$$

$$x_{1-R1-2} + x_{2-R1-2} + x_{3-R1-2} + x_{4-R1-2} + x_{5-R1-2} \leq -G_{21}; \tag{5.143}$$

$$x_{1-R1-3} + x_{2-R1-3} + x_{3-R1-3} + x_{4-R1-3} + x_{5-R1-3} \leq -G_{31}; \tag{5.144}$$

Table 5.8 Eliminated nodes for a two-region decomposition

132	143	295	148	255
133	144	298	149	256
134	145	301	252	257
135	146	304	253	258
136	147	307	254	259
260	261	271	272	273
274				
$G_{13}=0$	$G_{23}=0$	$G_{33}=0$	$G_{43}=0$	$G_{53}=0$

$$x_{1-R1-4}+x_{2-R1-4}+x_{3-R1-4}+x_{4-R1-4}+x_{5-R1-4}\leq -G_{41}; \tag{5.145}$$

$$x_{1-R1-5}+x_{2-R1-5}+x_{3-R1-5}+x_{4-R1-5}+x_{5-R1-5}\leq -G_{51}; \tag{5.146}$$

The nodal constraints remain the same except for removing the $R3$ variable. As seen in Table 5.8, certain nodes are not considered. The coefficients are assigned a zero if the variable is not involved with the decomposition process. Nodes B1 and B2 are not considered.

5.5 Formulating the IEEE 30-Bus System's Constraints

The Dantzig-Wolfe implementation is also tested with the next level of the IEEE bus system. The IEEE 30-bus system has 6 generators and 20 loads as shown in Fig. 5.10. This system is much larger compared to the IEEE 14-bus system that was discussed previously. The generator and load data are given in Tables 5.9 and 5.10, respectively.

The risk, or loss factor is calculated for objective-function coefficients using the failure-rate and repair-rate data shown in Table 5.11.

The IEEE network model for the 30-bus system is shown in Fig. 5.11, and the decomposed regions are shown in Fig. 5.12.

5.5.1 *Nodal Constraints for Region 1*

5.5.1.1 Node 1

$$x_{1-2-1}+x_{1-3-1}-(x_{2-1-1}+x_{3-1-1})+(x_{1-R1-1})\leq G_{11}; \tag{5.147}$$

$$x_{1-2-2}+x_{1-3-2}-(x_{2-1-2}+x_{3-1-2})+(x_{1-R1-2})\leq G_{21}; \tag{5.148}$$

$$x_{1-2-3}+x_{1-3-3}-(x_{2-1-3}+x_{3-1-3})+(x_{1-R1-3})\leq G_{31}; \tag{5.149}$$

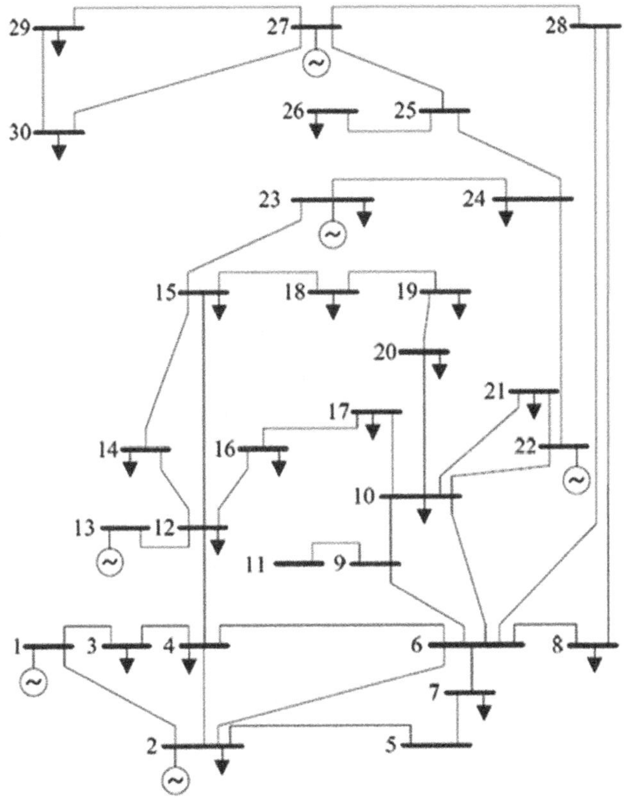

Fig. 5.10 IEEE 30-bus system single-line diagram

$$x_{1-2-4} + x_{1-3-4} - (x_{2-1-4} + x_{3-1-4}) + (x_{1-R1-4}) \leq G_{41}; \tag{5.150}$$

$$x_{1-2-5} + x_{1-3-5} - (x_{2-1-5} + x_{3-1-5}) + (x_{1-R1-5}) \leq G_{51}; \tag{5.151}$$

$$x_{1-2-6} + x_{1-3-6} - (x_{2-1-6} + x_{3-1-6}) + (x_{1-R1-6}) \leq G_{61}; \tag{5.152}$$

5.5.1.2 Node 2

$$\begin{aligned} & x_{2-4-1} + x_{2-5-1} + x_{2-6-1} + x_{2-1-1} \\ & \quad - (x_{4-2-1} + x_{5-2-1} + x_{6-2-1} + x_{1-2-1}) + (x_{2-R1-1}) \\ & \quad \leq G_{21} - 21.7; \end{aligned} \tag{5.153}$$

Table 5.9 Generator data for the IEEE 30-bus system [WG10]

Generator data of the IEEE 30-bus test system													
		Cost coefficients											
Unit	Bus	A ($/MWh^2)	B ($/MWh)	C ($)	P_{max} (MW)	P_{min} (MW)	Min up time (h)	Min down time (h)	Ramp up (MW)	Ramp Down (MW)	Startup Ramp (MW)	Shutdown ramp (MW)
Gl	1	0.0200	15.00	0	80	15	2	2	25	25	70	60
G2	2	0.0175	14.75	0	80	15	2	2	25	25	70	60
G3	13	0.0250	16.00	0	50	10	3	3	15	15	70	60
G4	22	0.0625	14.00	0	50	10	4	4	15	15	70	60
G5	23	0.0250	16.00	0	30	5	3	3	10	10	70	60
G6	27	0.0083	15.25	0	55	10	4	4	15	15	70	60

Table 5.10 Demand profile for the IEEE 30-bus system

Nodes	Load demand	Nodes	Load demand
1	0	16	3.5
2	21.7	17	9.0
3	2.4	18	3.2
4	67.6	19	9.5
5	34.2	20	2.2
6	0	21	17.5
7	22.8	22	0
8	30	23	3.2
9	0	24	8.7
10	5.8	25	0
11	0	26	3.5
12	11.2	27	0
13	0	28	0
14	6.2	29	2.4
15	8.2	30	10.6

$$x_{2-4-2} + x_{2-5-2} + x_{2-6-2} + x_{2-1-2} - (x_{4-2-2} + x_{5-2-2} + x_{6-2-2} + x_{1-2-2}) + (x_{2-R1-2}) \leq G_{22} - 21.7; \tag{5.154}$$

$$x_{2-4-3} + x_{2-5-3} + x_{2-6-3} + x_{2-1-3} - (x_{4-2-3} + x_{5-2-3} + x_{6-2-3} + x_{1-2-3}) + (x_{2-R1-3}) \leq G_{23} - 21.7; \tag{5.155}$$

$$x_{2-4-4} + x_{2-5-4} + x_{2-6-4} + x_{2-1-4} - (x_{4-2-4} + x_{5-2-4} + x_{6-2-4} + x_{1-2-4}) + (x_{2-R1-4}) \leq G_{24} - 21.7; \tag{5.156}$$

$$x_{2-4-5} + x_{2-5-5} + x_{2-6-5} + x_{2-1-5} - (x_{4-2-5} + x_{5-2-5} + x_{6-2-5} + x_{1-2-5}) + (x_{2-R1-5}) \leq G_{25} - 21.7; \tag{5.157}$$

$$x_{2-4-6} + x_{2-5-6} + x_{2-6-6} + x_{2-1-6} - (x_{4-2-6} + x_{5-2-6} + x_{6-2-6} + x_{1-2-6}) + (x_{2-R1-6}) \leq G_{26} - 21.7; \tag{5.158}$$

5.5.1.3 Node 3

$$x_{1-3-1} + x_{3-4-1} - (x_{3-1-1} + x_{4-3-1}) + (x_{3-R1-1}) \leq G_{11} - 2.4; \tag{5.159}$$

Table 5.11 Repair and failure rates for the IEEE 30-bus system

Network data of the IEEE 30-bus test system						
Line No.	From	To	X (p.u.)	Flow limit (MW)	Failure rate	Repair rate
1	1	2	0.06	130	0.9783	0.0217
2	1	3	0.19	130	0.9841	0 0159
3	2	4	0.17	65	0.9532	0.0468
4	3	4	0.04	130	0.9172	0.0828
5	2	5	0.20	130	0.9786	0.0214
6	2	6	0.18	65	0.9497	0.0503
7	4	6	0.04	90	0.9828	0.0172
8	5	7	0.12	70	0.9760	0.0240
9	6	7	0.08	130	0.9211	0 0789
10	6	8	0.04	32	0.9494	0.0506
11	6	9	0.21	65	0.9494	0.0506
12	6	10	0.56	32	0.9211	0.0789
13	9	11	0.21	65	0.9535	0.0465
14	9	10	0.11	65	0.9509	0.0491
15	4	12	0.26	65	0.9660	0.0340
16	12	13	0.14	65	0.9838	0.0162
17	12	14	0.26	32	0.9754	0 0246
18	12	15	0.13	32	0.9598	0.0402
19	12	16	0.20	32	0.9510	0.0490
20	14	15	0.20	16	0.9494	0.0506
21	16	17	0.19	16	0.9494	0.0506
22	15	18	0.22	16	0.9236	0.0764
23	18	19	0.13	16	0.9514	0.0486
24	19	20	0.07	32	0.9509	0.0491
25	10	20	0.21	32	0.9666	0.0334
26	10	17	0.08	32	0.9824	0.0176
27	10	21	0.07	32	0.9786	0.0214
28	10	22	0.15	32	0.9612	0.0388
29	21	22	0.02	32	0.9462	0.0538
30	15	23	0.20	16	0.9498	0 0502
31	22	24	0.18	16	0.9506	0.0494
32	23	24	0.27	16	0.9181	0.0819
33	24	25	0.33	16	0.9483	0.0517
34	25	25	0.38	16	0.9537	0.0463
35	25	27	0.21	16	0.9733	0.0267
36	28	27	0.40	65	0.9818	0.0182
37	27	29	0.42	16	0.9808	0.0192
38	27	30	0.60	16	0.9564	0.0436
39	29	30	0.45	16	0.9537	0.0463
40	8	28	0.20	32	0.9537	0.0463
41	6	28	0.06	32	0.9536	0.0464

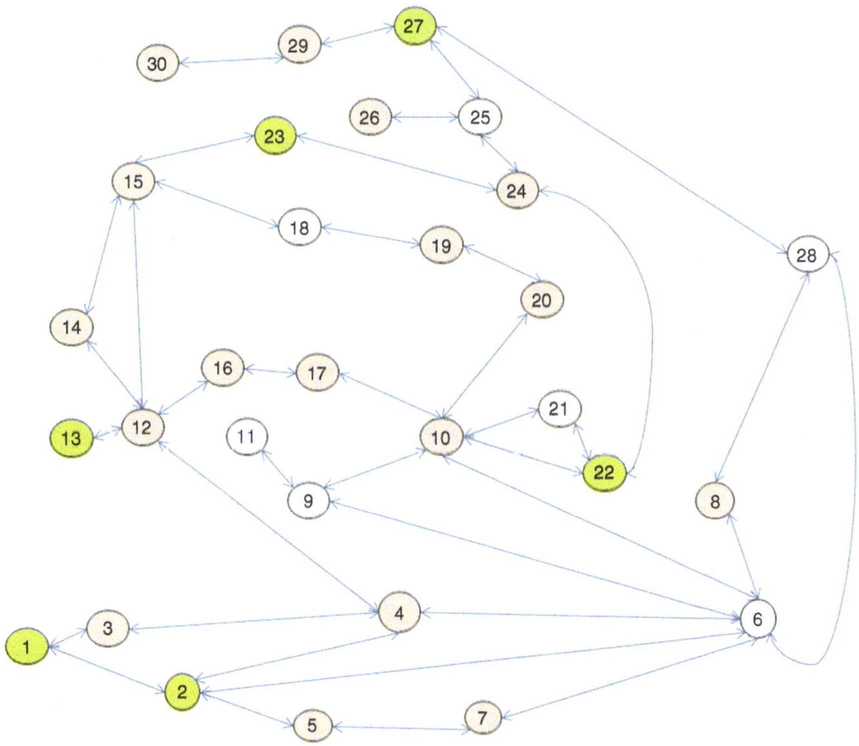

Fig. 5.11 Network model for the IEEE 30-bus system

$$x_{1-3-2} + x_{3-4-2} - (x_{3-1-2} + x_{4-3-2}) + (x_{3-R1-2}) \leq G_{21} - 2.4; \quad (5.160)$$

$$x_{1-3-3} + x_{3-4-3} - (x_{3-1-3} + x_{4-3-3}) + (x_{3-R1-3}) \leq G_{31} - 2.4; \quad (5.161)$$

$$x_{1-3-4} + x_{3-4-4} - (x_{3-1-4} + x_{4-3-4}) + (x_{3-R1-4}) \leq G_{41} - 2.4; \quad (5.162)$$

$$x_{1-3-5} + x_{3-4-5} - (x_{3-1-5} + x_{4-3-5}) + (x_{3-R1-5}) \leq G_{51} - 2.4; \quad (5.163)$$

$$x_{1-3-6} + x_{3-4-6} - (x_{3-1-6} + x_{4-3-6}) + (x_{3-R1-6}) \leq G_{61} - 2.4; \quad (5.164)$$

5.5.1.4 Node 4

$$\begin{aligned} &x_{4-6-1} + x_{4-3-1} + x_{4-2-1} + x_{4-12-1} \\ &\quad - (x_{6-4-1} + x_{3-4-1} + x_{2-4-1} + x_{12-4-1}) + (x_{4-R1-1}) \\ &\quad \leq G_{11} - 67.6; \end{aligned} \quad (5.165)$$

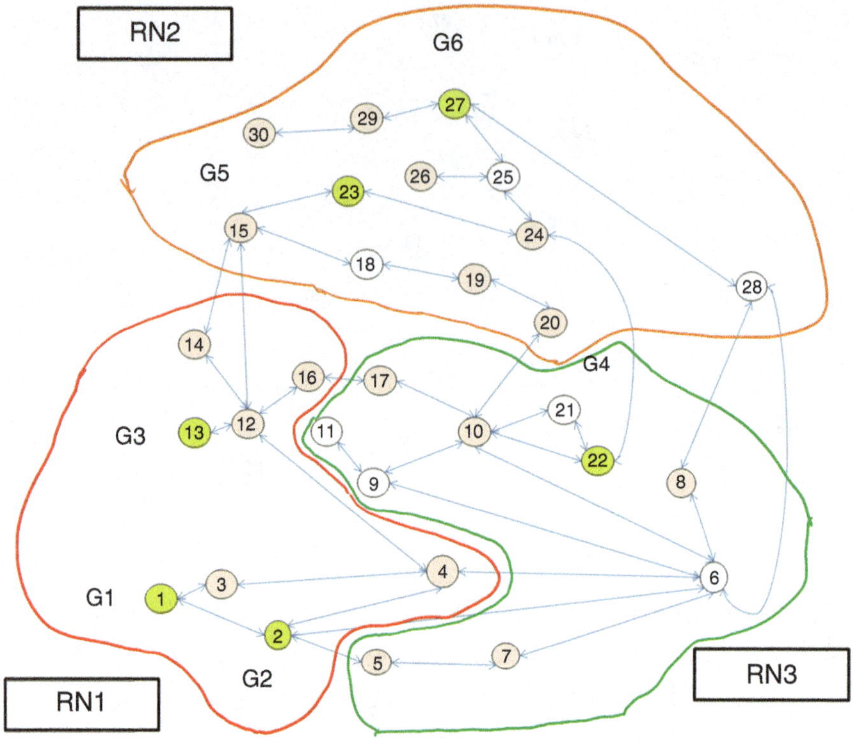

Fig. 5.12 Network model for decomposing the IEEE 30-bus system to three regions

$$x_{4-6-2} + x_{4-3-2} + x_{4-2-2} + x_{4-12-2} - (x_{6-4-2} + x_{3-4-2} + x_{2-4-2} + x_{12-4-2}) + (x_{4-R1-2}) \leq G_{21} - 67.6; \tag{5.166}$$

$$x_{4-6-3} + x_{4-3-3} + x_{4-2-3} + x_{4-12-3} - (x_{6-4-3} + x_{3-4-3} + x_{2-4-3} + x_{12-4-3}) + (x_{4-R1-3}) \leq G_{31} - 67.6; \tag{5.167}$$

$$x_{4-6-4} + x_{4-3-4} + x_{4-2-4} + x_{4-42-4} - (x_{6-4-4} + x_{3-4-4} + x_{2-4-4} + x_{42-4-4}) + (x_{4-R1-4}) \leq G_{41} - 67.6; \tag{5.168}$$

$$x_{4-6-5} + x_{4-3-5} + x_{4-2-5} + x_{4-12-5} - (x_{6-4-5} + x_{3-4-5} + x_{2-4-5} + x_{12-4-5}) + (x_{4-R1-5}) \leq G_{51} - 67.6; \tag{5.169}$$

$$\begin{aligned} & x_{4-6-6} + x_{4-3-6} + x_{4-2-6} + x_{4-12-6} \\ & \quad - (x_{6-4-6} + x_{3-4-6} + x_{2-4-6} + x_{12-4-6}) + (x_{4-R1-6}) \\ & \quad \leq G_{61} - 67.6; \end{aligned} \tag{5.170}$$

5.5.1.5 Node 12

$$\begin{aligned} & x_{12-13-1} + x_{12-4-1} + x_{12-16-1} + x_{12-14-1} + x_{12-15-1} \\ & \quad - (x_{13-12-1} + x_{4-12-1} + x_{16-12-1} + x_{14-12-1} + x_{15-12-1}) \\ & \quad + (x_{12-R1-1}) \\ & \quad \leq G_{11} - 11.2; \end{aligned} \tag{5.171}$$

$$\begin{aligned} & x_{12-13-2} + x_{12-4-2} + x_{12-16-2} + x_{12-14-2} + x_{12-15-2} \\ & \quad - (x_{13-12-2} + x_{4-12-2} + x_{16-12-2} + x_{14-12-2} + x_{15-12-2}) \\ & \quad + (x_{12-R1-2}) \\ & \quad \leq G_{21} - 11.2; \end{aligned} \tag{5.172}$$

$$\begin{aligned} & x_{12-13-3} + x_{12-4-3} + x_{12-16-3} + x_{12-14-3} + x_{12-15-3} \\ & \quad - (x_{13-12-3} + x_{4-12-3} + x_{16-12-3} + x_{14-12-3} + x_{15-12-3}) \\ & \quad + (x_{12-R1-3}) \\ & \quad \leq G_{31} - 11.2; \end{aligned} \tag{5.173}$$

$$\begin{aligned} & x_{12-13-4} + x_{12-4-4} + x_{12-16-4} + x_{12-14-4} + x_{12-15-4} \\ & \quad - (x_{13-12-4} + x_{4-12-4} + x_{16-12-4} + x_{14-12-4} + x_{15-12-4}) \\ & \quad + (x_{12-R1-4}) \\ & \quad \leq G_{41} - 11.2; \end{aligned} \tag{5.174}$$

$$\begin{aligned} & x_{12-13-5} + x_{12-4-5} + x_{12-16-5} + x_{12-14-5} + x_{12-15-5} \\ & \quad - (x_{13-12-5} + x_{4-12-5} + x_{16-12-5} + x_{14-12-5} + x_{15-12-5}) \\ & \quad + (x_{12-R1-5}) \\ & \quad \leq G_{51} - 11.2; \end{aligned} \tag{5.175}$$

$$\begin{aligned} & x_{12-13-6} + x_{12-4-6} + x_{12-16-6} + x_{12-14-6} + x_{12-15-6} \\ & \quad - (x_{13-12-6} + x_{4-12-6} + x_{16-12-6} + x_{14-12-6} + x_{15-12-6}) \\ & \quad + (x_{12-R1-6}) \\ & \quad \leq G_{61} - 11.2; \end{aligned} \tag{5.176}$$

5.5.1.6 Node 13

$$x_{13-12-1} - (x_{12-13-1}) + (x_{13-R1-1}) \leq 0; \tag{5.177}$$

$$x_{13-12-2} - (x_{12-13-2}) + (x_{13-R1-2}) \leq 0; \tag{5.178}$$

$$x_{13-12-3} - (x_{12-13-3}) + (x_{13-R1-3}) \leq 0; \quad (5.179)$$
$$x_{13-12-4} - (x_{12-13-4}) + (x_{13-R1-4}) \leq 0; \quad (5.180)$$
$$x_{13-12-5} - (x_{12-13-5}) + (x_{13-R1-5}) \leq 0; \quad (5.181)$$
$$x_{13-12-6} - (x_{12-13-6}) + (x_{13-R1-6}) \leq 0; \quad (5.182)$$

5.5.1.7 Node 14

$$x_{14-15-1} + x_{14-12-1} - (x_{12-14-1} + x_{15-14-1}) + (x_{14-R1-1}) \leq 0; \quad (5.183)$$
$$x_{14-15-2} + x_{14-12-2} - (x_{12-14-2} + x_{15-14-2}) + (x_{14-R1-2}) \leq 0; \quad (5.184)$$
$$x_{14-15-3} + x_{14-12-3} - (x_{12-14-3} + x_{15-14-3}) + (x_{14-R1-3}) \leq 0; \quad (5.185)$$
$$x_{14-15-4} + x_{14-12-4} - (x_{12-14-1} + x_{15-14-4}) + (x_{14-R1-4}) \leq 0; \quad (5.186)$$
$$x_{14-15-5} + x_{14-12-5} - (x_{12-14-5} + x_{15-14-5}) + (x_{14-R1-5}) \leq 0; \quad (5.187)$$
$$x_{14-15-6} + x_{14-12-6} - (x_{12-14-6} + x_{15-14-6}) + (x_{14-R1-6}) \leq 0; \quad (5.188)$$

5.5.1.8 Node 16

$$x_{16-12-1} + x_{16-17-1} - (x_{12-16-1} + x_{17-16-1}) + (x_{16-R1-1}) \leq G_{11} - 3.5; \quad (5.189)$$
$$x_{16-12-2} + x_{16-17-2} - (x_{12-16-2} + x_{17-16-2}) + (x_{16-R1-2}) \leq G_{21} - 3.5; \quad (5.190)$$
$$x_{16-12-3} + x_{16-17-3} - (x_{12-16-3} + x_{17-16-3}) + (x_{16-R1-3}) \leq G_{31} - 3.5; \quad (5.191)$$
$$x_{16-12-4} + x_{16-17-4} - (x_{12-16-4} + x_{17-16-4}) + (x_{16-R1-4}) \leq G_{41} - 3.5; \quad (5.192)$$
$$x_{16-12-5} + x_{16-17-5} - (x_{12-16-5} + x_{17-16-5}) + (x_{16-R1-5}) \leq G_{51} - 3.5; \quad (5.193)$$
$$x_{16-12-6} + x_{16-17-6} - (x_{12-16-6} + x_{17-16-6}) + (x_{16-R1-6}) \leq G_{61} - 3.5; \quad (5.194)$$

5.5.1.9 Artificial Nodes in Region 1

The nodes that participate in region 1 are nodes 1, 2, 3, 4, 5, 12, 13, 14, and 16. Each node is connected to $R1$ nodes.

$$x_{1-R1-1} + x_{2-R1-1} + x_{3-R1-1} + x_{4-R1-1} + x_{5-R1-1} + x_{12-R1-1} + x_{13-R1-1} + x_{14-R1-1} + x_{16-R1-1} \leq -G_{11}; \quad (5.195)$$

$$x_{1-R1-2} + x_{2-R1-2} + x_{3-R1-2} + x_{4-R1-2} + x_{5-R1-2} + x_{12-R1-2} + x_{13-R1-2} + x_{14-R1-2} + x_{16-R1-2} \leq -G_{21}; \quad (5.196)$$

$$x_{1-R1-3} + x_{2-R1-3} + x_{3-R1-3} + x_{4-R1-3} + x_{5-R1-3} + x_{12-R1-3} + x_{13-R1-3} + x_{14-R1-3} + x_{16-R1-3} \leq -G_{31}; \quad (5.197)$$

$$x_{1-R1-4} + x_{2-R1-4} + x_{3-R1-4} + x_{4-R1-4} + x_{5-R1-4} + x_{12-R1-4} + x_{13-R1-4} + x_{14-R1-4} + x_{16-R1-4} \leq -G_{41}; \quad (5.198)$$

$$x_{1-R1-5} + x_{2-R1-5} + x_{3-R1-5} + x_{4-R1-5} + x_{5-R1-5} + x_{12-R1-5} + x_{13-R1-5} + x_{14-R1-5} + x_{16-R1-5} \leq -G_{51}; \quad (5.199)$$

$$x_{1-R1-6} + x_{2-R1-6} + x_{3-R1-6} + x_{4-R1-6} + x_{5-R1-6} + x_{12-R1-6} + x_{13-R1-6} + x_{14-R1-6} + x_{16-R1-6} \leq -G_{61}; \quad (5.200)$$

5.5.2 *Nodal Constraints for Region 2*

5.5.2.1 Node 30

$$x_{30-29-1} - (x_{29-30-1}) + (x_{30-R2-1}) \leq G_{12} - 10.6; \quad (5.201)$$

$$x_{30-29-2} - (x_{29-30-2}) + (x_{30-R2-2}) \leq G_{22} - 10.6; \quad (5.202)$$

$$x_{30-29-3} - (x_{29-30-3}) + (x_{30-R2-3}) \leq G_{32} - 10.6; \quad (5.203)$$

$$x_{30-29-4} - (x_{29-30-4}) + (x_{30-R2-4}) \leq G_{42} - 10.6; \quad (5.204)$$

$$x_{30-29-5} - (x_{29-30-5}) + (x_{30-R2-5}) \leq G_{52} - 10.6; \quad (5.205)$$

$$x_{30-29-6} - (x_{29-30-6}) + (x_{30-R2-6}) \leq G_{62} - 10.6; \quad (5.206)$$

5.5.2.2 Node 29

$$x_{29-30-1} + x_{29-27-1} - (x_{30-29-1} + x_{27-29-1}) + (x_{29-R2-1}) \leq G_{12} - 2.4; \quad (5.207)$$

$$x_{29-30-2} + x_{29-27-2} - (x_{30-29-2} + x_{27-29-2}) + (x_{29-R2-2}) \leq G_{22} - 2.4; \quad (5.208)$$

$$x_{29-30-3} + x_{29-27-3} - (x_{30-29-3} + x_{27-29-3}) + (x_{29-R2-3}) \leq G_{32} - 2.4; \quad (5.209)$$

$$x_{29-30-4} + x_{29-27-4} - (x_{30-29-4} + x_{27-29-4}) + (x_{29-R2-4}) \leq G_{42} - 2.4; \quad (5.210)$$

$$x_{29-30-5} + x_{29-27-5} - (x_{30-29-5} + x_{27-29-5}) + (x_{29-R2-5}) \leq G_{52} - 2.4; \quad (5.211)$$

$$x_{29-30-6} + x_{29-27-6} - (x_{30-29-6} + x_{27-29-6}) + (x_{29-R2-6}) \leq G_{62} - 2.4; \quad (5.212)$$

5.5.2.3 Node 27

$$x_{27-29-1} + x_{27-25-1} - (x_{29-27-1} + x_{25-27-1}) + (x_{27-R2-1}) \leq G_{12}; \quad (5.213)$$

$$x_{27-29-2} + x_{27-25-2} - (x_{29-27-2} + x_{25-27-2}) + (x_{27-R2-2}) \leq G_{22}; \quad (5.214)$$

$$x_{27-29-3} + x_{27-25-3} - (x_{29-27-3} + x_{25-27-3}) + (x_{27-R2-3}) \leq G_{32}; \quad (5.215)$$

$$x_{27-29-4} + x_{27-25-4} - (x_{29-27-4} + x_{25-27-4}) + (x_{27-R2-4}) \leq G_{42}; \quad (5.216)$$

$$x_{27-29-5} + x_{27-25-5} - (x_{29-27-5} + x_{25-27-5}) + (x_{27-R2-5}) \leq G_{52}; \quad (5.217)$$

$$x_{27-29-6} + x_{27-25-6} - (x_{29-27-6} + x_{25-27-6}) + (x_{27-R2-6}) \leq G_{62}; \quad (5.218)$$

5.5.2.4 Node 15

$$\begin{aligned} &x_{15-14-1} + x_{15-12-1} + x_{15-18-1} + x_{15-23-1} \\ &\quad - (x_{14-15-1} + x_{12-15-1} + x_{18-15-1} + x_{23-15-1}) + (x_{15-R2-1}) \\ &\quad \leq G_{12} - 8.2; \end{aligned} \quad (5.219)$$

$$\begin{aligned} &x_{15-14-2} + x_{15-12-2} + x_{15-18-2} + x_{15-23-2} \\ &\quad - (x_{14-15-2} + x_{12-15-2} + x_{18-15-2} + x_{23-15-2}) + (x_{15-R2-2}) \\ &\quad \leq G_{22} - 8.2; \end{aligned} \quad (5.220)$$

$$\begin{aligned} &x_{15-14-3} + x_{15-12-3} + x_{15-18-3} + x_{15-23-3} \\ &\quad - (x_{14-15-3} + x_{12-15-3} + x_{18-15-3} + x_{23-15-3}) + (x_{15-R2-3}) \\ &\quad \leq G_{32} - 8.2; \end{aligned} \quad (5.221)$$

$$\begin{aligned} &x_{15-14-4} + x_{15-12-4} + x_{15-18-4} + x_{15-23-4} \\ &\quad - (x_{14-15-4} + x_{12-15-4} + x_{18-15-4} + x_{23-15-4}) + (x_{15-R2-4}) \\ &\quad \leq G_{42} - 8.2; \end{aligned} \quad (5.222)$$

$$\begin{aligned} &x_{15-14-5} + x_{15-12-5} + x_{15-18-5} + x_{15-23-5} \\ &\quad - (x_{14-15-5} + x_{12-15-5} + x_{18-15-5} + x_{23-15-5}) + (x_{15-R2-5}) \\ &\quad \leq G_{52} - 8.2; \end{aligned} \quad (5.223)$$

$$x_{15-14-6} + x_{15-12-6} + x_{15-18-6} + x_{15-23-6} - (x_{14-15-6} + x_{12-15-6} + x_{18-15-6} + x_{23-15-6}) + (x_{15-R2-6}) \leq G_{62} - 8.2; \quad (5.224)$$

Similarly, flow-balance constraints for nodes 23, 26, 18, 19, 20, 24, 25, and 28 can be formulated. Due to space constraints, we have not included them here.

$$x_{28-27-1} + x_{28-8-1} + x_{28-6-1} - (x_{27-28-1} + x_{8-28-1} + x_{6-28-1}) + (x_{28-R2-1}) \leq G_{12}; \quad (5.225)$$

$$x_{28-27-2} + x_{28-8-2} + x_{28-6-2} - (x_{27-28-2} + x_{8-28-2} + x_{6-28-2}) + (x_{28-R2-2}) \leq G_{22}; \quad (5.226)$$

$$x_{28-27-3} + x_{28-8-3} + x_{28-6-3} - (x_{27-28-3} + x_{8-28-3} + x_{6-28-3}) + (x_{28-R2-3}) \leq G_{32}; \quad (5.227)$$

$$x_{28-27-4} + x_{28-8-4} + x_{28-6-4} - (x_{27-28-4} + x_{8-28-4} + x_{6-28-4}) + (x_{28-R2-4}) \leq G_{42}; \quad (5.228)$$

$$x_{28-27-5} + x_{28-8-5} + x_{28-6-5} - (x_{27-28-5} + x_{8-28-5} + x_{6-28-5}) + (x_{28-R2-5}) \leq G_{52}; \quad (5.229)$$

$$x_{28-27-6} + x_{28-8-6} + x_{28-6-6} - (x_{27-28-6} + x_{8-28-6} + x_{6-28-6}) + (x_{28-R2-6}) \leq G_{62}; \quad (5.230)$$

5.5.2.5 Joint-Capacity Constraints for Nodes in Region 2

$$x_{30-R2-1} + x_{29-R2-1} + x_{27-R2-1} + x_{15-R2-1} + x_{23-R2-1} + x_{26-R2-1} + x_{18-R2-1} + x_{19-R2-1} + x_{20-R2-1} + x_{24-R2-1} + x_{25-R2-1} + x_{28-R2-1} \leq -G_{12}; \quad (5.231)$$

$$x_{30-R2-1} + x_{29-R2-1} + x_{27-R2-1} + x_{15-R2-1} + x_{23-R2-1} + x_{26-R2-1} + x_{18-R2-1} + x_{19-R2-1} + x_{20-R2-1} + x_{24-R2-1} + x_{25-R2-1} + x_{28-R2-1} \leq -G_{22}; \quad (5.232)$$

$$x_{30-R2-1} + x_{29-R2-1} + x_{27-R2-1} + x_{15-R2-1} + x_{23-R2-1} + x_{26-R2-1} + x_{18-R2-1} + x_{19-R2-1} + x_{20-R2-1} + x_{24-R2-1} + x_{25-R2-1} + x_{28-R2-1} \leq -G_{32}; \quad (5.233)$$

$$x_{30-R2-1} + x_{29-R2-1} + x_{27-R2-1} + x_{15-R2-1} + x_{23-R2-1} + x_{26-R2-1} + x_{18-R2-1} + x_{19-R2-1} + x_{20-R2-1} + x_{24-R2-1} + x_{25-R2-1} + x_{28-R2-1} \leq -G_{42}; \quad (5.234)$$

$$x_{30-R2-1} + x_{29-R2-1} + x_{27-R2-1} + x_{15-R2-1} + x_{23-R2-1} + x_{26-R2-1} + x_{18-R2-1} + x_{19-R2-1} + x_{20-R2-1} + x_{24-R2-1} + x_{25-R2-1} + x_{28-R2-1} \leq -G_{52}; \quad (5.235)$$

$$x_{30-R2-1} + x_{29-R2-1} + x_{27-R2-1} + x_{15-R2-1} + x_{23-R2-1} + x_{26-R2-1} + x_{18-R2-1} + x_{19-R2-1} + x_{20-R2-1} + x_{24-R2-1} + x_{25-R2-1} + x_{28-R2-1} \leq -G_{62}; \quad (5.236)$$

5.5.3 *Nodal Constraints for Region 3*

The nodes that participate in region 3 are nodes 5, 7, 6, 8, 22, 21, 10, 17, 11, and 9.

5.5.3.1 Node 5

$$x_{5-7-1} + x_{5-2-1} - (x_{7-5-1} + x_{2-5-1}) + (x_{5-R3-1}) \leq G_{13} - 34.2; \quad (5.237)$$

$$x_{5-7-2} + x_{5-2-2} - (x_{7-5-2} + x_{2-5-2}) + (x_{5-R3-2}) \leq G_{23} - 34.2; \quad (5.238)$$

$$x_{5-7-3} + x_{5-2-3} - (x_{7-5-3} + x_{2-5-3}) + (x_{5-R3-3}) \leq G_{33} - 34.2; \quad (5.239)$$

$$x_{5-7-4} + x_{5-2-4} - (x_{7-5-4} + x_{2-5-4}) + (x_{5-R3-4}) \leq G_{43} - 34.2; \quad (5.240)$$

$$x_{5-7-5} + x_{5-2-5} - (x_{7-5-5} + x_{2-5-5}) + (x_{5-R3-5}) \leq G_{53} - 34.2; \quad (5.241)$$

$$x_{5-7-6} + x_{5-2-6} - (x_{7-5-6} + x_{2-5-6}) + (x_{5-R3-6}) \leq G_{63} - 34.2; \quad (5.242)$$

Similarly, constraints for nodes 7, 6, 8, 22, 21, 10, 17, 11, and 9 can be formulated.

5.5.3.2 Node 9

$$x_{9-11-1} + x_{9-10-1} + x_{9-6-1} - (x_{11-9-1} + x_{10-9-1} + x_{6-9-1}) + (x_{9-R1-1}) \leq G_{13}; \quad (5.243)$$

$$x_{9-11-2} + x_{9-10-2} + x_{9-6-2} - (x_{11-9-2} + x_{10-9-2} + x_{6-9-2}) + (x_{9-R1-2}) \leq G_{23}; \quad (5.244)$$

$$x_{9-11-3} + x_{9-10-3} + x_{9-6-3} - (x_{11-9-3} + x_{10-9-3} + x_{6-9-3}) + (x_{9-R1-3}) \leq G_{33}; \quad (5.245)$$

$$x_{9-11-4} + x_{9-10-4} + x_{9-6-4} - (x_{11-9-4} + x_{10-9-4} + x_{6-9-4}) + (x_{9-R1-4}) \leq G_{43}; \tag{5.246}$$

$$x_{9-11-5} + x_{9-10-5} + x_{9-6-5} - (x_{11-9-5} + x_{10-9-5} + x_{6-9-5}) + (x_{9-R1-5}) \leq G_{53}; \tag{5.247}$$

$$x_{9-11-6} + x_{9-10-6} + x_{9-6-6} - (x_{11-9-6} + x_{10-9-6} + x_{6-9-6}) + (x_{9-R1-6}) \leq G_{63}; \tag{5.248}$$

5.5.3.3 Joint-Capacity Constraints for Region 3

$$x_{5-R3-1} + x_{7-R3-1} + x_{6-R3-1} + x_{8-R3-1} + x_{22-R3-1} + x_{21-R3-1} + x_{10-R3-1} + x_{17-R3-1} + x_{11-R3-1} + x_{9-R3-1} \leq -G_{13}; \tag{5.249}$$

$$x_{5-R3-2} + x_{7-R3-2} + x_{6-R3-2} + x_{8-R3-2} + x_{22-R3-2} + x_{21-R3-2} + x_{10-R3-2} + x_{17-R3-2} + x_{11-R3-2} + x_{9-R3-2} \leq -G_{23}; \tag{5.250}$$

$$x_{5-R3-3} + x_{7-R3-3} + x_{6-R3-3} + x_{8-R3-3} + x_{22-R3-3} + x_{21-R3-3} + x_{10-R3-3} + x_{17-R3-3} + x_{11-R3-3} + x_{9-R3-3} \leq -G_{33}; \tag{5.251}$$

$$x_{5-R3-4} + x_{7-R3-4} + x_{6-R3-4} + x_{8-R3-4} + x_{22-R3-4} + x_{21-R3-4} + x_{10-R3-4} + x_{17-R3-4} + x_{11-R3-4} + x_{9-R3-4} \leq -G_{43}; \tag{5.252}$$

$$x_{5-R3-5} + x_{7-R3-5} + x_{6-R3-5} + x_{8-R3-5} + x_{22-R3-5} + x_{21-R3-5} + x_{10-R3-5} + x_{17-R3-5} + x_{11-R3-5} + x_{9-R3-5} \leq -G_{53}; \tag{5.253}$$

$$x_{5-R3-6} + x_{7-R3-6} + x_{6-R3-6} + x_{8-R3-6} + x_{22-R3-6} + x_{21-R3-6} + x_{10-R3-6} + x_{17-R3-6} + x_{11-R3-6} + x_{9-R3-6} \leq -G_{63}; \tag{5.254}$$

The additional constraints that interconnect regions

$$x_{a1,14,1} - (x_{14,c1,1}) = G_{12}; \tag{5.255}$$

$$x_{a1,14,2} - (x_{14,c1,2}) = G_{22}; \tag{5.256}$$

$$x_{a1,14,3} - (x_{14,c1,3}) = G_{32}; \tag{5.257}$$

$$x_{a1,14,4} - (x_{14,c1,1}) = G_{42}; \tag{5.258}$$

$$x_{a1,14,5} - (x_{14,c1,5}) = G_{52}; \tag{5.259}$$

$$x_{a1,14,6} - (x_{14,c1,6}) = G_{62}; \tag{5.260}$$

$$x_{a2,5,1} - (x_{5,b1,1}) = G_{13}; \tag{5.261}$$

$$x_{a2,5,2} - (x_{5,b1,2}) = G_{23}; \tag{5.262}$$

$$x_{a2,5,3} - (x_{5,b1,3}) = G_{33}; \tag{5.263}$$

$$x_{a2,5,4} - (x_{5,b1,4}) = G_{43}; \tag{5.264}$$

$$x_{a2,5,5} - (x_{5,b1,5}) = G_{53}; \tag{5.265}$$

$$x_{a2,5,6} - (x_{5,b1,6}) = G_{63}; \tag{5.266}$$

$$x_{a2,a1,1} = G_{11}; \tag{5.267}$$

$$x_{a2,a1,2} = G_{21}; \tag{5.268}$$

$$x_{a2,a1,3} = G_{31}; \tag{5.269}$$

$$x_{a2,a1,4} = G_{41}; \tag{5.270}$$

$$x_{a2,a1,5} = G_{51}; \tag{5.271}$$

$$x_{a2,a1,6} = G_{61}; \tag{5.272}$$

and additional constraints.

Thus, the LP formulation for the IEEE 14-bus and 30-bus systems is developed. Now, we investigate the AMPL implementation of the Dantzig-Wolfe procedure and the results in Chap. 6.

Chapter 6
Implementation and Testing of the Dantzig-Wolfe Procedure

This chapter discusses the AMPL implementation and results for running the IEEE 14-bus and IEEE 30-bus systems. The environment for the AMPL modeling software is discussed regarding how to specify the model, data, and run-file information.

6.1 Overview of Modeling in AMPL and the Results

Practical, large-scale mathematical programming involves more than just the minimization or maximization of an objective function subject to constraint equations and inequalities. Before any optimizing algorithm can be applied, some effort must be expended to formulate the underlying model and to generate the requisite computational data structures. If algorithms could deal with optimization problems as people do, then the formulation and generation phases of modeling might be relatively easy. In reality, however, there are many differences between the form in which human modelers understand a problem and the form in which algorithms solve it. Reliable translation from the "modeler's form to the algorithm's form" is often a considerable expense.

In the traditional approach for translation, the work is divided between a human and a computer. First, a person who understands the modeler's form writes a computer program where the output represents the required data structures. Then, a computer compiles and executes the program to create the algorithm's form. This arrangement is often costly and error-prone; most seriously, the program must be debugged by a human modeler even though the algorithm's output form is not meant for people to read.

In the important special case of linear programming, the largest part of the algorithm's form is representing the constraint's coefficient matrix. Typically, this matrix is a very sparse one where rows and columns number in the hundreds or thousands, and where nonzero elements appear in intricate patterns. A computer

P. Ranganathan, K.E. Nygard, *Distributed Linear Programming Models in a Smart Grid*, Power Electronics and Power Systems, DOI 10.1007/978-3-319-52617-1_6

program that produces a compact representation of the coefficients is called a matrix generator.

Compared to previous languages, AMPL is notable for the generality of its syntax and for the similarity of its expressions to the algebraic notation customarily used in the modeler's form. AMPL offers a variety of types and operations to define indexing sets as well as a range of logical expressions. AMPL draws considerable inspiration from the XML prototype language [Fou83], incorporating many changes and extensions.

AMPL is a new language that is designed to make these steps easier and less error-prone. AMPL closely resembles the symbolic algebraic notation that many modelers use to describe mathematical programs, yet it is regular and formal enough to be processed by a computer system; it is particularly notable for the generality of its syntax and for the variety of its indexing operations. We have implemented a translator that takes a linear AMPL model and the associated data as input and then produces output that is suitable for standard Dantzig-Wolfe linear-programming optimizers.

6.2 Lagrangian Relaxation Procedure

Dual decomposition, and more generally Lagrangian relaxation, is a classical method for combinatorial optimization [Sal04]. Dual decomposition leverages the observation that many decoding problems can be broken into two or more sub-problems, together with linear constraints that enforce some notion of agreement among the different problems' solutions. The sub-problems are chosen such that they can be solved efficiently by using exact combinatorial algorithms. The agreement constraints are incorporated using Lagrange multipliers, and an iterative algorithm—for example, a sub-gradient algorithm—is used to minimize the resulting dual variables. Dual decomposition algorithms have the following properties. They are typically simple and efficient. For example, sub-gradient algorithms involve two steps for each iteration: first, each sub-problem is solved using a combinatorial algorithm; second, simple additive updates are made to the Lagrange multipliers. They have well-understood formal properties, particularly through connections to linear-programming (LP) relaxations. In cases where the underlying LP relaxation is tight, they produce an exact solution for the original decoding problem, with a certificate of optimality. In cases where the underlying LP is not tight, heuristic methods can be used to derive a good solution; alternatively, constraints can be added incrementally until the relaxation is tight, at which point an exact solution is recovered.

Dual decomposition, where two or more combinatorial algorithms are used, is a special case of Lagrangian relaxation (LR). It is useful to consider LR methods that utilize a single combinatorial algorithm together with a set of linear constraints that are, again, incorporated using Lagrange multipliers. Utilizing a single combinatorial algorithm is qualitatively different from dual-decomposition approaches, although the techniques are very closely related. Lagrangian relaxation has a long history in the combinatorial-optimization literature, going back to the seminal 1971 work of Held and Karp who derived a relaxation algorithm for the traveling-salesman problem.

The Lagrangian relaxation of general LP is given as

$$Z = \min Cx \text{ subject to } Ax \leq b, Bx \leq d \text{ and } x \geq 0$$

The DW version of the LP in equation is given by

$$\text{Minimize} Z = cx - \lambda^k (Ax - b_o) \tag{6.1}$$

subject to

$$Bx \leq b; \quad x_i \geq 0; \quad b_i \geq 0 \tag{6.2}$$

Figure 6.1 illustrates the interaction between the sub-problems and master problems via dual variables and theta. The Lagrangian multiplier is chosen iteratively by the AMPL code to solve the problem.

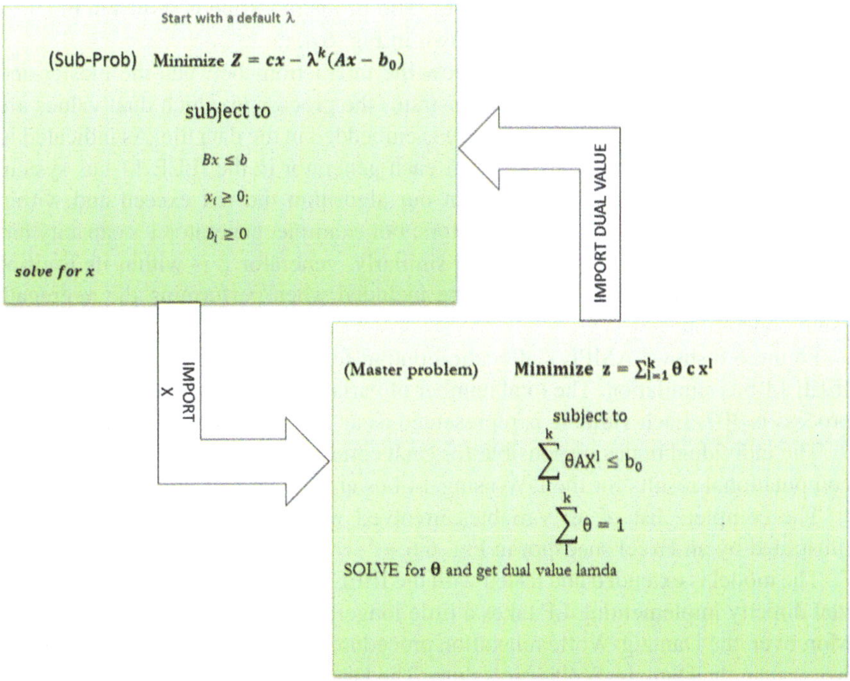

Fig. 6.1 Lagrangian relaxation of the Dantzig-Wolfe decomposition

6.3 Computational Results of IEEE Bus System

In testing this approach, we observe that the Lagrangian relaxation version of the DW performs poorly when compared to other models such as Direct LP and Direct DW implementation.

Table 6.1 indicates the real power values of the loads and generators for an IEEE 14-bus system. There is a balance in the total supply and demand for this system which equals 258 MW. This system has five generators and 11 loads as seen in Table 6.1. The initial assumption about generator values to individual regions is shown in Table 6.2.

To run the DW procedure, we make initial assumptions about the generator values as shown in Table 6.2.. These assumptions are reasonable when considering the number of loads and generators in the given system. A print-screen view of the model, data, and run file is shown in Fig. 6.2.

The number of master constraints or complicating constraints is indicated using the "param" command in AMPL. For example, the "cr" variable refers to master constraints, and the "or" variable refers to region constraints or other constraints. The sub-problem classification is determined using the "nsub" parameter. The convexity constraint is denoted using the lambda variable. The sub-problem matrices are denoted using the *d* and *f* variables in the model file. Separate data and run files are used to run the program. AMPL's sub-problem setting is shown in Fig. 6.3, and the commodity constraints are shown in Fig. 6.4.

The model file in Fig. 6.3 shows how the interactions between the master and sub-problems occur iteratively and illustrates the process in which dual values are calculated. For convenience, the run file is embedded in the data file. As indicated in Fig. 6.4, the commodity constraints for each generator in the IEEE 14-bus system are satisfied. This figure indicates that our algorithm did not exceed and within limits of the capacities of these generators. For example, generator 1's capacity has not exceeded more than 88 MW, and similarly, generator 2 is within its limit of 60 MW. The nodes that participate are included when performing the aggregate "sum" operation.

Figure 6.5 shows AMPL's allocation output for all variables involved with the IEEE 14-bus simulation. The total number of variables involved with the allocation process is 307. Each variable is represented as a "node" in the AMPL modeling.

The individual nodes responsible for each commodity are listed in Fig. 6.6. The computational results for the DW using 14 bus are shown in Table 6.3.

The complete list of all variables involved with the IEEE 14-bus system is illustrated by an Excel snapshot in Fig. 6.6.

The model is extended and tested with the IEEE 30-bus system. The results show that directly implementing LP takes a little longer than the decomposition scheme. Moreover, the Dantzig-Wolfe relaxation procedure takes much longer and performs worse than the Dantzig-Wolfe procedure. The Dantzig-Wolfe implementation runs faster and provides reasonable computational time savings. Table 6.3 illustrates the 14-bus system's performances with multi-region decomposition. The total number

Table 6.1 Supplies and the demand profile for the IEEE 14-bus system

	L1(2)	L2(3)	L3(5)	L4(4)	L5(6)	L6(11)	L7(10)	L8(9)	L9(12)	L10(13)	L11(14)	Supply
G1 (1)	0.55	1.7	.14	4.47	.15	.16	.52	.11	.424	7.67	8.21	88
G2 (2)	1	0.15	6.34	9.92	6.35	6.36	6.72	7.31	6.624	6.87	7.41	60
G3 (3)	6.15	1	12.79	8.53	15.88	15.89	16.25	16.84	16.15	16.4	16.9	60
G4 (8)	11.92	10.53	9.34	2.0	3.31	3.3	2.94	2.35	3.58	3.83	3.33	25
G5 (6)	6.35	15.88	0.01	7.35	1	0.01	0.37	0.96	0.274	0.44	0.98	25
	21.7	94.2	7.6	47	11.2	3.5	9	29.5	6	13.5	14.8	258

Table 6.2 Initial allocation of generator values for three regions

Generators	Region 1	Region 2	Region 3
G1	58.10	19.36	9.68
G2	39.65	13.20	6.60
G3	39.65	13.20	6.60
G4	16.50	5.50	2.75
G5	16.52	5.50	2.75

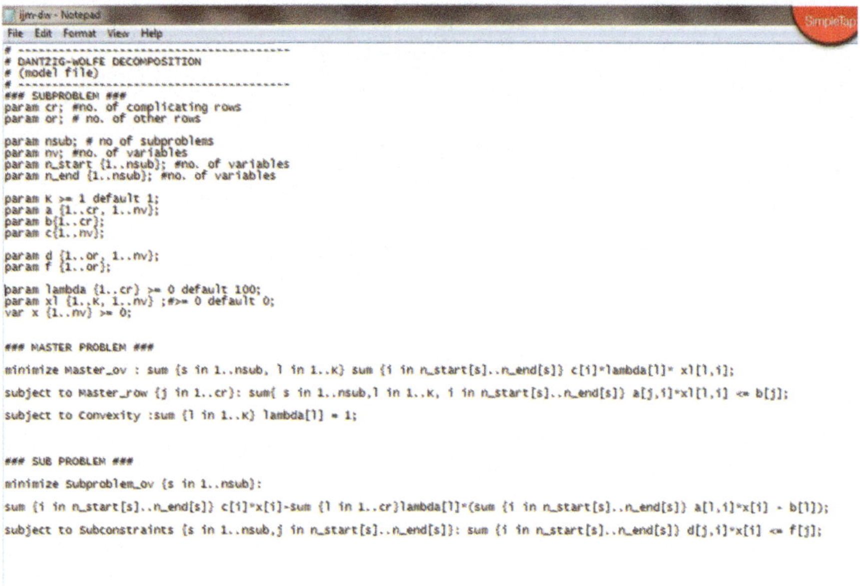


```
# ----------------------------------------
# DANTZIG-WOLFE DECOMPOSITION
# (model file)
# ----------------------------------------
### SUBPROBLEM ###
param cr; #no. of complicating rows
param or; # no. of other rows

param nsub; # no of subproblems
param nv; #no. of variables
param n_start {1..nsub}; #no. of variables
param n_end {1..nsub}; #no. of variables

param K >= 1 default 1;
param a {1..cr, 1..nv};
param b{1..cr};
param c{1..nv};

param d {1..or, 1..nv};
param f {1..or};

param lambda {1..cr} >= 0 default 100;
param xl {1..K, 1..nv} ;#>= 0 default 0;
var x {1..nv} >= 0;


### MASTER PROBLEM ###

minimize Master_ov : sum {s in 1..nsub, l in 1..K} sum {i in n_start[s]..n_end[s]} c[i]*lambda[l]* xl[l,i];

subject to Master_row {j in 1..cr}: sum{ s in 1..nsub,l in 1..K, i in n_start[s]..n_end[s]} a[j,i]*xl[l,i] <= b[j];

subject to Convexity :sum {l in 1..K} lambda[l] = 1;


### SUB PROBLEM ###

minimize Subproblem_ov {s in 1..nsub}:

sum {i in n_start[s]..n_end[s]} c[i]*x[i]-sum {l in 1..cr}lambda[l]*(sum {i in n_start[s]..n_end[s]} a[l,i]*x[i] - b[l]);

subject to Subconstraints {s in 1..nsub,j in n_start[s]..n_end[s]}: sum {i in n_start[s]..n_end[s]} d[j,i]*x[i] <= f[j];
```

Fig. 6.2 Snapshot of the AMPL model file that shows the DW implementation

of variables in the 14-bus and 30-bus systems is 307 and 650 variables with 130 and 225 sub-constraints, respectively. The computational results for the IEEE 30-bus system with various decomposition structures are given in Table 6.4. The allocation and objective value for the cost parameter yields in same solution. The interactions between the master and sub-problems take 212 iterations to attain an optimal cost for the three-region decomposition. The same formulation can be broken into two regions with some modifications. Certain variables that link to nodes are not considered. The coefficients are assigned as zero if the variable is not part of the decomposition process. For example, joint-capacity constraint R3 is not involved with the two-region problem. Similarly, nodes B1 and B2 are neither considered nor removed.

Our results show huge savings regarding computational cost and response time for the entire IEEE 30-bus system compared to the Direct LP and Lagrangian relaxation formulations. The key contribution is that we have developed, implemented, and tested the Dantzig-Wolfe procedure in the IEEE bus system

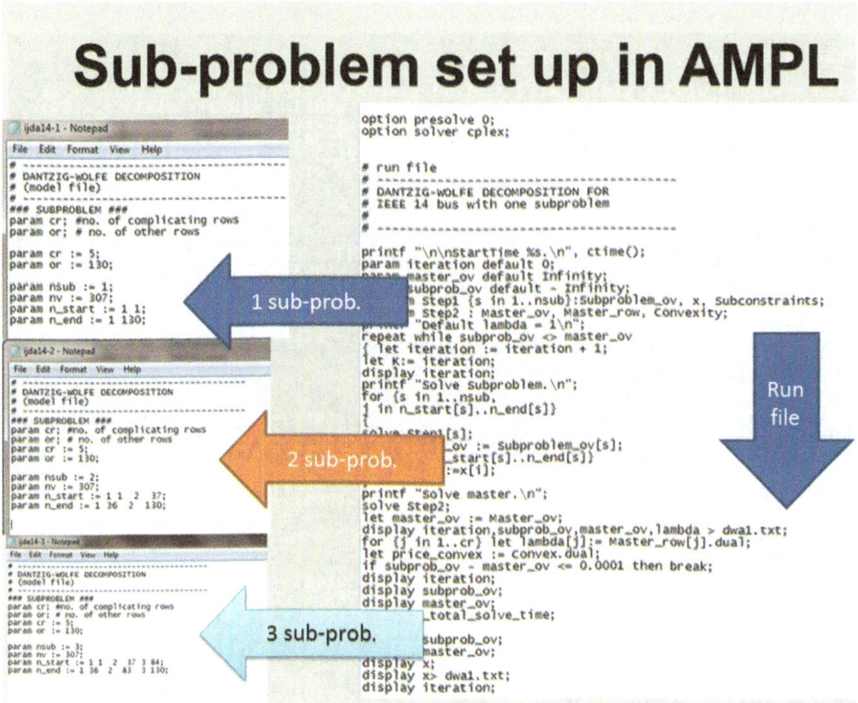

Fig. 6.3 AMPL model file of the DW

with various decomposition structures. It is important to note that all decomposition results and methods result in the same cost and allocation values, the main contribution of this book.

Sensitivity analysis of the IEEE 14-bus system and the IEEE 30-bus system for loss (failures and repair rates) is investigated. Finding the optimal solution for a linear-programming model is important, but it is not the only information available. There is a tremendous amount of sensitivity information, or information about what happens when data values are changed. When formulating a problem as a linear program, we have to invoke a certainty assumption: we have to know what values the data took; finally, decisions are made based on the data from the LP run. Often, this assumption is somewhat dubious: the data might be unknown, guessed, or otherwise inaccurate. How can we determine the effect on the optimal decisions if values such as the failure or repair rates change? Clearly, some numbers in the data are more important than others. Can we find the “important” numbers? Can we determine the effect of misestimating? Linear programming offers extensive capabilities to address these questions. In the model, we test the sensitivity of our LP formulation with respect to the effect on line failures. We have simulated certain line failures by treating certain variables to zero and noticing the change in the allocation procedure.

Fig. 6.4 Snapshot of the commodity constraints

In an IEEE 14-bus system, we simulated line failures in Region 1 by treating x_{12}, x_{23} and x_{24} to 0. The authors were able to observe the re-allocation of power to loads via other lines such as x_{15} and x_{54} to other regions and eventually reaching an optimum value. Similarly, in an IEEE 30-bus system, lines x_{25}, x_{57} and x_{76} were set to zero, and the allocation's re-routing can be seen via the x_{13}, x_{34} lines to other regions. Thus, the failure scenario can easily be modeled by treating those lines as zero. Therefore, it is evident from these AMPL runs that the sensitivity of the DW allocation process and the transmission lines for the IEEE 14-bus and IEEE 30-bus systems are bound to the flow limits set on the transmission lines.

The inferences about the results and possible future tasks related to this procedure are discussed in the final chapter with conclusions.

x [*] :=

1	79.7893	78	0	155	0	232	0.729333
2	-52.0857	79	0	156	0	233	0
3	0	80	0	157	0	234	0
4	0	81	0	158	0	235	0.744513
5	-27.7036	82	0	159	0	236	0.733333
6	101.849	83	0	160	0	237	0
7	-52.0857	84	0	161	-21.1407	238	0
8	22.06	85	0	162	4.75549	239	0
9	0	86	0	163	0	240	0
10	-27.7036	87	0	164	0	241	0
11	79.7893	88	0	165	0	242	0
12	-52.0857	89	0	166	0	243	0
13	0	90	0	167	0	244	0
14	0	91	0	168	4.75549	245	0
15	-27.7036	92	0	169	0	246	0
16	79.7893	93	0	170	0	247	0
17	-52.0857	94	0	171	0	248	0.733566
18	0	95	0	172	0	249	0.733566
19	0	96	0	173	0	250	0
20	-27.7036	97	0	174	0	251	0
21	79.7893	98	0	175	0	252	0
22	-52.0857	99	0	176	0	253	0
23	0	100	0	177	36.4289	254	0
24	0	101	-4.245	178	0	255	-0.0335394
25	-27.7036	102	27.7371	179	0	256	0
26	0	103	0	180	0	257	0
27	19.2964	104	-4.245	181	0	258	0
28	66.4964	105	0	182	36.4289	259	0
29	0	106	-4.245	183	0	260	0
30	0	107	0	184	0	261	0
31	0	108	-4.245	185	0	262	0
32	0	109	0	186	0	263	0
33	0	110	-4.245	187	0	264	0
34	19.2964	111	0	188	0	265	0
35	66.4964	112	-15.445	189	0	266	0
36	0	113	-29.935	190	0	267	0
37	0	114	-15.445	191	0	268	0
38	0	115	-29.935	192	-14.712	269	0
39	0	116	-15.445	193	-14.6993	270	0
40	0	117	-29.935	194	-14.6993	271	0
41	41.3564	118	-15.445	195	-14.7335	272	0
42	66.4964	119	-29.935	196	-14.7	273	0
43	0	120	-15.445	197	-15.5413	274	0
44	22.06	121	-29.935	198	-15.5329	275	0
45	0	122	0	199	-15.5329	276	0
46	0	123	0	200	-15.511	277	0
47	0	124	0	201	-15.5333	278	0
48	19.2964	125	0	202	0	279	0
49	66.4964	126	0	203	0	280	0
50	0	127	0	204	0	281	0
51	0	128	0	205	0	282	0
52	0	129	0	206	0	283	0
53	0	130	0	207	0	284	0
54	0	131	0	208	0	285	0
55	19.2964	132	4.245	209	0	286	0
56	71.4233	133	4.245	210	0	287	0
57	0	134	4.245	211	0	288	0
58	0	135	4.245	212	0	289	0
59	4.92691	136	4.245	213	0	290	0
60	0	137	-47.68	214	0	291	0
61	0	138	0	215	0	292	0
62	0	139	0	216	0	293	0.476826
63	0	140	0	217	0.0120015	294	-0.0382921
64	0	141	0	218	-0.000697857	295	-0.438534
65	0	142	0	219	-0.000697857	296	0.698703
66	0	143	0	220	0	297	-0.0555203
67	0	144	0	221	0	298	-0.643183
68	0	145	0	222	4.24133	299	0.698703
69	0	146	0	223	4.96643	300	-0.0555203
70	0	147	0	224	4.96643	301	-0.643183
71	0	148	52.4355	225	4.24451	302	1.679
72	0	149	52.4355	226	4.23333	303	-0.135366
73	0	150	-21.1407	227	0	304	-1.54364
74	0	151	-21.1407	228	0	305	1.67697
75	0	152	-21.1407	229	0	306	-0.133333
76	0	153	0	230	0	307	-1.54364
77	0	154	0	231	0		

Fig. 6.5 Allocation results of the IEEE 14-bus system

1	X-1-2-1	51	X4-2-4	101	x5,a2,1	151	xa1-6-4	201	x14-9-5	251	x11-R2-5	301	G33
2	X1-5-1	52	X3-2-4	102	xa2,5,4	152	xa1-a2-4	202	x9-10-1	252	x14-b2-1	302	G41
3	X2-1-1	53	X2-R1-4	103	x5-R1-1	153	xa2-a1-1	203	x9-10-2	253	x14-b2-2	303	G42
4	X5-1-1	54	X2-5-5	104	x5,a2,2	154	xa1-6-1	204	x9-10-3	254	x14-b2-3	304	G43
5	X1-R1-1	55	X2-4-5	105	x5-R1-2	155	xa2-a1-2	205	x9-10-4	255	x14-b2-4	305	G51
6	X1-2-2	56	X2-3-5	106	x5,a2,3	156	xa1-6-2	206	x9-10-5	256	x14-b2-5	306	G52
7	X1-5-2	57	X5-2-5	107	x5-R1-3	157	x5-a2-1	207	x9-14-1	257	xb2-14-1	307	G53
8	X2-1-2	58	X4-2-5	108	x5,a2,4	158	x5-a2-2	208	x9-14-2	258	xb2-14-2		
9	X5-1-2	59	X3-2-5	109	x5-R1-4	159	x5-a2-3	209	x9-14-3	259	xb2-14-3		
10	X1-R1-2	60	X2-R1-5	110	x5,a2,5	160	xa2-a1-3	210	x9-14-4	260	xb2-14-4		
11	X1-2-3	61	X3-4-1	111	x5-R1-5	161	xa2-5-4	211	x9-14-5	261	xb2-14-5		
12	X1-5-3	62	X2-3-1	112	x6-12-1	162	xb1-b2-4	212	x9-c2-1	262	x14-R2-1		
13	X2-1-3	63	X4-3-1	113	x12-13-1	163	xb2-b1-5	213	x9-c2-2	263	x14-R2-2		
14	X5-1-3	64	X3-R1-1	114	x6-12-2	164	xb1-6-5	214	x9-c2-3	264	x14-R2-3		
15	X1-R1-3	65	X3-4-2	115	x12-13-2	165	xb1-13-5	215	x9-c2-4	265	x14-R2-4		
16	X1-2-4	66	X2-3-2	116	x6-12-3	166	x14-b2-5	216	x9-c2-5	266	x14-R2-5		
17	X1-5-4	67	X4-3-2	117	x12-13-3	167	x11-b2-5	217	x9-R2-1	267	x7-R2-4		
18	X2-1-4	68	X3-R1-2	118	x6-12-4	168	xb2-14-4	218	x9-R2-2	268	x7-R2-5		
19	X5-1-4	69	X3-4-3	119	x12-13-4	169	xb2-11-4	219	x9-R2-3	269	x13-b1-4		
20	X1-R1-4	70	X2-3-3	120	x6-12-5	170	xc2-7-1	220	x9-R2-4	270	x6-b1-4		
21	X1-2-5	71	X4-3-3	121	x12-13-5	171	xc2-7-2	221	x9-R2-5	271	xb1-b2-4		
22	X1-5-5	72	X3-R1-3	122	x12-6-1	172	xc2-7-3	222	x10-11-1	272	xb2-b1-5		
23	X2-1-5	73	X3-4-4	123	x12-6-2	173	x7-c2-5	223	x10-11-2	273	xb1-6-5		
24	X5-1-5	74	X2-3-4	124	x12-6-3	174	x7-R2-1	224	x10-11-3	274	xb1-13-5		
25	X1-R1-5	75	X4-3-4	125	x12-6-4	175	x7-R2-2	225	x10-11-4	275	x4-c1-1		
26	x2-5-1	76	X3-R1-4	126	x12-6-5	176	x7-R2-3	226	x10-11-5	276	x4-c1-2		
27	x2-4-1	77	X3-4-5	127	x13-12-1	177	x8-7-1	227	x11-10-1	277	x4-c1-3		
28	x2-3-1	78	X2-3-5	128	x13-12-2	178	x8-7-2	228	x11-10-2	278	x4-c1-4		
29	x5-2-1	79	X4-3-5	129	x13-12-3	179	x8-7-3	229	x11-10-3	279	x4-c1-5		
30	x4-2-1	80	X3-R1-5	130	x13-12-4	180	x8-7-4	230	x11-10-4	280	xc1-c2-1		
31	x3-2-1	81	X4-5-1	131	x13-12-5	181	x8-7-5	231	x11-10-5	281	xc1-c2-2		
32	x2-R1-1	82	X4-3-1	132	x12-R3-1	182	x7-8-1	232	x10-R2-1	282	xc1-c2-3		
33	X2-5-2	83	X5-4-1	133	x12-R3-2	183	x7-8-2	233	x10-R2-2	283	xc1-c2-4		
34	X2-4-2	84	X4-R1-1	134	x12-R3-3	184	x7-8-3	234	x10-R2-3	284	xc1-c2-5		
35	X2-3-2	85	X4-5-2	135	x12-R3-4	185	x7-8-4	235	x10-R2-4	285	x9-c2-4		
36	X5-2-2	86	X4-3-2	136	x12-R3-5	186	x7-8-5	236	x10-R2-5	286	xc2-c1-4		
37	X4-2-2	87	X5-4-2	137	X13-B1-4	187	x8-R2-1	237	x11-b2-1	287	x7-c2-4		
38	X3-2-2	88	X4-R1-2	138	x13-R3-1	188	x8-R2-2	238	x11-b2-2	288	x9-c2-5		
39	X2-R1-2	89	X4-5-3	139	x13-R3-2	189	x8-R2-3	239	x11-b2-3	289	xc2-c1-5		
40	X2-5-3	90	X4-3-3	140	x13-R3-3	190	x8-R2-4	240	x11-b2-4	290	xc2-9-1		
41	X2-4-3	91	X5-4-3	141	x13-R3-4	191	x8-R2-5	241	x11-b2-5	291	xc2-9-2		
42	X2-3-3	92	X4-R1-3	142	x13-R3-5	192	X10-9-1	242	xb2-11-1	292	xc2-9-3		
43	X5-2-3	93	X4-5-4	143	x6-R3-1	193	X10-9-2	243	xb2-11-2	293	G11		
44	X4-2-3	94	X4-3-4	144	x6-R3-2	194	x10-9-3	244	xb2-11-3	294	G12		
45	X3-2-3	95	X5-4-4	145	x6-R3-3	195	x10-9-4	245	xb2-11-4	295	G13		
46	X2-R1-3	96	X4-R1-4	146	x6-R3-4	196	x10-9-5	246	xb2-11-5	296	G21		
47	X2-5-4	97	X4-5-5	147	x6-R3-5	197	x14-9-1	247	x11-R2-1	297	G22		
48	X-2-4-4	98	X4-3-5	148	x6-B1-4	198	x14-9-2	248	x11-R2-2	298	G23		
49	X2-3-4	99	X5-4-5	149	xb1-6-4	199	x14-9-3	249	x11-R2-3	299	G31		
50	X5-2-4	100	X4-R1-5	150	x6-a1-4	200	x14-9-4	250	x11-R2-4	300	G32		

Fig. 6.6 Snapshot of the nodes and variables in the IEEE 14-bus simulation

Table 6.3 Computational results for the IEEE 14-bus system

Parameters	Direct LP	Lag with 1 sub-prob. (Relax.)	Lag with 2 sub-prob. (Relax.)	Lag with 3 sub-prob. (Relax.)	Lag with 10 sub-prob. (Relax.)	DW with 1 sub-prob.	DW with 2 sub-prob.	DW with 3 sub-prob.	DW with 10 sub-prob.
Computation time (s)	1.5	4.781	4.64	4.6875	4.796	2.437	2.421	2.328	2.375
Variables	307	307	307	307	307	307	307	307	307
Total # of constraints	135	SP 1: 130	SP 1:1–36	SP 1:1–36	SP 1: 1–20	SP 1: 125	SP 1:1–36	SP 1:1–36	SP 1: 1–20
					SP 2: 21–40				SP 2: 21–40
					SP 3: 41–50				SP 3: 41–50
					SP 4: 51–60				SP 4: 51–60
									SP 5: 61–80
					SP 5: 61–80		SP 2:37–130	SP 2:37–83	SP 6: 81–90
SP: Sub-problem			SP 2:37–130	SP 2: 37–83	SP 6: 81–90				SP 7: 91–100
					SP 7: 91–100				SP 8: 101–110
					SP 8: 101–100	MP: 5	MP: 5	SP 3:84–130	SP 9: 111–115
					SP 9: 111–115				SP10:116–130
MC: Master constraints		MP: 5	MP: 5	SP 3: 84–130	SP10:116–130			MP: 5	MP: 5
				MP: 5	MP: 5				
Cost	14,015	14,015	14,015	14,015	14,015	14,015	14,015	14,015	14,015
Allocation result	Same	Same	Same	Same	Same	Same	Same	Same	Same

Table 6.4 Computational results for the IEEE 30-bus system

Parameters	Direct LP	Lag with 1 sub-prob. (Relax.)	Lag with 2 sub-prob. (Relax.)	Lag with 3 sub-prob. (Relax.)	Lag with 10 sub-prob. (Relax.)	DW with 1 sub-prob.	DW with 2 sub-prob.	DW with 3 sub-prob.	DW with 10 sub-prob.
Computation time: (s)	4.5	14.781	14.765	14.6875	14.96	3.4375	3.421	3.38	3.95
Variables	650	650	650	650	650	650	650	650	650
Total # of constraints	275	SP 1: 270	SP 1:1–36	SP 1: 1–36 SP 2:37–150	SP 1: 1–20 SP 2: 21–40 SP 3: 41–50 SP 4: 51–60	SP 1: 270	SP 1: 1–36 SP 2: 37–275	SP 1:1–36 SP 2:37–83	SP 1: 1–20 SP 2: 21–40 SP 3: 41–50 SP 4: 51–60 SP 5: 61–80
SP: Sub-problem			SP 2:37–275	SP 3:150–275	SP 5: 61–80 SP 6: 81–90 SP 7: 91–100 SP 8: 101–110 SP 9: 111–115 SP10:116–275	MP: 5	MP: 5	SP 3:84–275 MP: 5	SP 6: 81–90 SP 7: 91–100 SP 8: 101–110 SP 9: 111–115 SP10:116–275
MC: Master constraints		MP: 5	MP: 5	MP: 5	MP: 5				MP: 5
Cost	67485.4	67485.4	67485.4	67485.4	67485.4	67485.4	67485.4	67485.4	67485.4
Allocation result	Same	Same	Same	Same	Same	Same	Same	Same	Same

Chapter 7
Remarks About the Dantzig-Wolfe Scheme

A distributed linear-programming model has been created, developed, implemented, and tested. Two standard IEEE bus systems are modeled and successfully decomposed, in multiple ways, into sub-problems. The problem is solved iteratively in each case and directly supports resource allocation in a Smart-grid environment. I have shown that the LP-based design using the Dantzig-Wolfe decomposition can execute and can quickly determine the primary resource scheduling and allocation issues if a failure occurs in the grid. The decomposition procedure can easily be managed by system operators. In the study using the 4-bus, 14-bus, and 30-bus systems, the results indicate that the computational benefits of the Dantzig-Wolfe approach enable fast responses on the order of a millisecond to a few seconds as network size increases. Although the 30-bus system is not a large bus network, the results clearly indicate a faster computation time if an appropriate Dantzig-Wolfe structure is formulated. This approach can enable system operators in the electric grid to respond to any allocation request for resources in the event of outages or line failures. The book's key contribution is the design, development, and testing of a procedure that successfully decomposes an optimization problem that is defined over a large grid but can be solved in regional pieces.

The following inferences are made for my defined problem:

Inference 1: The larger size of decomposing into regions does not guarantee computational savings for the overall problem. For example, the computational time savings is greater with the three-region decomposition of the IEEE network rather than with the ten-region decomposition network, an important and interesting contribution of this book because it conveys that not all decompositions can yield computation-time savings. However, the solution procedure is decomposed by regions, which significantly spreads out the computational load. In addition, the procedure demonstrates that feasible resource-allocation solutions can be obtained on an intermediate basis, allowing solutions to be terminated early with a still-valuable heuristic problem solution.

P. Ranganathan, K.E. Nygard, *Distributed Linear Programming Models in a Smart Grid*, Power Electronics and Power Systems, DOI 10.1007/978-3-319-52617-1_7

Inference 2: The Dantzig-Wolfe decomposition performs better with the IEEE 30-bus system than the IEEE 14-bus system. This finding is probably due to increased variables as well as the large number of constraints and many inter-related, complicated constraints with the sub-problems in the tested network's structure. Also, a large number of iterations are required between the sub-problems and the master model.

Inference 3: The Lagrangian-Relaxation procedure performs poorly compared to the actual Dantzig-Wolfe version. This finding suggests that a more sophisticated procedure for setting Lagrangian multipliers is need.

Inference 4: Direct LP formulation performs better with the IEEE 14-bus system compared to the decomposition scheme. This result is likely due to the relatively small size of the test problems.

Inference 5: All models have identical resource allocation and identical cost as measured by the objective-function value. This demonstrates that the computational procedures, although solved through decomposition, still provide the best possible solution.

Inference 6: The number of iterations for interactions between the master problem and sub-problems is different, ranging from 100 to 300 iterations. The approach takes multiple iterations to reach an optimal cost as the solution for our objective, which demonstrates the appropriate and accurate interactions between the master and sub-problem constraints.

Inference 7: The proposed DW method is tested for scalability up to the IEEE 30-bus system. Due to the large number of constraints for a given formulation, scalability issues remain.

Inference 8: The benefits and significance of decomposition by regions yields reasonable time savings for the computations compared to running all constraint sets directly.

In summary, the key contribution is that we have developed, implemented, and tested the Dantzig-Wolfe procedure in the IEEE bus system with various decomposition structures. It is important to note that all decomposition structures result in the same cost and allocation values. This is a significant and main contribution for this book. In addition, I have modeled the LP formulation for resource allocation with the known uncertainty information included in Chap. 3. A branch-and-bound based algorithm for resource allocation is also presented in Chap. 4 as part of the contribution.

In this book, we define scalability as the ability of a DW process to handle a growing number of sub-problems or constraints in a capable manner, or its ability to be enlarged to accommodate that growth. For example, it can refer to the capability of a system to increase the total throughput (such as computational time) under an increased load when resource constraints are added.

The DW algorithm is said to scale if it is suitably efficient and practical when applied to large problems, such as when there are a large number of participating nodes, as is the case with a typical, distributed Smart-grid system. The book shows that feasible scalability up to the level of the IEEE 30-bus system with 650 variables and 325 constraints.

The application of the evaluated decompositions can be implemented on a hierarchical US electric grid. For example, the practicality of a true large-scale grid can be envisioned as distributing resources among the four Independent System Operators (ISOs): the NYISO, MISO, Western Interconnection, and Southern Interconnection systems.

Regarding future plans for individuals who wish to study this problem, the authors recommend that they test using large-scale systems such as the standard 118-bus, 300-bus, and 1000-bus models. The efficiency of the Dantzig-Wolfe procedure relies on how complicated constraints are greatly involved with the sub-problem variables and solutions.

Chapter 8
A Linear Classifier for Decision Support in a Smart Grid

8.1 Introduction

As more electric utility companies move toward a smart-grid environment on a daily basis with the integration of renewables and the installation of a growing number of smart meters, researchers should address the problem of how to mine and manage the growing data that result from the grid's smart meters and other intelligent devices. In fact, some experts predict that the Smart-Grid (SG) data-analytics market will reach $4.2 billion by 2015 and that SG distribution automation spending will be $46 billion worldwide by 2015. As a result, there is enormous research and business potential in the data-analytics market [PR].

It has been studied recently in [SV10, SVY10] that the future for an efficient Smart Grid should explore ways to maximize the utilization of renewable energy sources, such as using GVs for sustainable Cyber Physical Energy Systems (CPES), in order to reduce emissions and cost. The chapter uses CPES data that were previously defined [SVY10].

Figure 8.1 shows a three-phase decision framework in a power system where data from multiple sources are collected in phase 1 through renewable sources, generators, sensors, and loads. Then, model tree-based decision models (M5 and J48) are applied after it is integrated in phase 2, and finally in phase 3, the classification or prediction of tree outcomes is distributed to the respective decision makers, agents, or units. In data preprocessing, the data selection, cleaning, and transformation are done. The data-preprocessing stage is where irrelevant or noisy data are identified and removed, and relevant data are extracted from the raw data [BF96]. There are many well-established approaches that deal with missing attributes or ambiguous responses, ranging from the most common attribute, the event-covering method, to ignoring the value altogether [GH01, LAP06].

We refer Power System operators (PSO) as personnel who act as a load dispatcher, substation inspector, power switchboard operator, hydro-electric station operator, electrical technician, or supervising engineer. PSOs monitor and operate

P. Ranganathan, K.E. Nygard, *Distributed Linear Programming Models in a Smart Grid*, Power Electronics and Power Systems, DOI 10.1007/978-3-319-52617-1_8

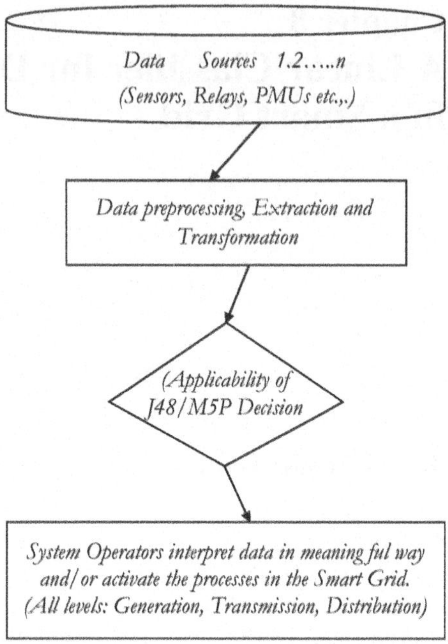

Fig. 8.1 Decision framework for the power-system operators

equipment on transmission lines and at transformer and power-generating stations. PSOs control the distribution and regulate the flow of electrical power in the transmission network. They are employed by electric-power generation, transmission, and distribution companies. They control the electricity that is generated and distributed to a particular region by monitoring and operating the computerized or pneumatic switchboards that regulate the power flow.

The distribution demands on a generating station or sub-station change daily due to system outages, repair work, and other factors. It is their role to coordinate, schedule, and direct power loads and line voltages to meet these demands. At any given time, there is repair work or construction that requires part of the system to be isolated and shut down. They consult operating drawings of the power system and prepare switching orders that will isolate the work areas without causing a power outage. They also have to consider voltages, load transfers, and line capacities in order not to overload part of the system. How do they make decisions to control such vast and complex grid operations? What computer-aided interface will help make this decision making easier or control the various processes?

In this chapter, we restrict our attention to the decision-making process for these operators with tree-based models using the available data from a Unit Commitment (UC) Problem. Because the power system's load varies throughout the day and reaches a different peak value from 1 day to another, the electric utility has to decide, in advance, which generators to start and when to connect them to the network, as well as the sequence in which the operating units should be shut down and for how long, based on the demand levels. The computational procedure to make such decisions is the unit commitment problem, and we say that a unit is committed when it is scheduled for connection to the power system.

8.2 Background on a Data Source from a GV

The source of data in this chapter is renewables, hence our approach to a Smart Grid has the following capabilities: (a) renewable energy sources, mainly wind and solar, are used to reduce the electric industry's emissions; (b) GVs are used to reduce the transportation industry's emissions; and (c) GVs are smartly used as loads; energy storages; and small, portable power plants.

As discussed [SVY10], the next-generation plug-in vehicles, including plug-in hybrid electric vehicles (PHEVs) and electric vehicles (EVs) with vehicle-to-grid capability, which are referred to as "gridable vehicles" (GVs) [SVY10], can reduce emissions for the transportation industry. GVs can be used as loads, energy sources (small, portable power plants) and energy storages in a smart grid that is integrated with renewable energy. Smart charging and discharging of the GVs in a distributed-energy source and load environment requires intelligent scheduling mechanisms which have a great potential for efficiently transporting electric energy.

Vehicle-to-grid (V2G) describes a system where plug-in electric vehicles, such as electric cars and plug-in hybrids (PHEVs), communicate with the power grid to sell demand-response services either by delivering electricity to the grid or by throttling the rates charged. Vehicle-to-grid can be used with such *gridable* vehicles, that is, plug-in electric vehicles (PHEVs) with grid capacity. Because most vehicles are parked an average of 90 plus percent of the time, their batteries could be used to let electricity flow from the car to the power lines and back, thereby saving costs and emissions. The concept allows V2G vehicles to provide power in order to help balance loads by "valley filling" (charging at night when demand is low) and "peak shaving" (sending power back to the grid when demand is high). It can give utilities new ways to provide regulation services (keeping voltage and frequency stable) and to provide spinning reserves (meet sudden demands for power). These V2G/G2V data are presented as one column in the unit commitment problem that we plan to analyze in the following sections (Fig. 8.2).

In the future development of Smart Grids, utilizing electric vehicles could buffer renewable power sources, such as wind or solar power, by storing excess energy that is produced during windy or sunny periods and then providing it back to the grid during high-load periods, thus effectively stabilizing the intermittency of wind and solar power. Some see this application of vehicle-to-grid technology as a renewable-energy approach that can penetrate the baseline electric market. This approach may even help public utilities to not have to build as many natural-gas or coal-fired power plants in order to meet peak demand. This chapter focuses on how data analytics can help with decision support for the generation, transmission, and distribution pieces of the grid. We tested our decision models (specifically the J48, decision-stump, and M4 models) using the WEKA data-mining tool from a 10-unit

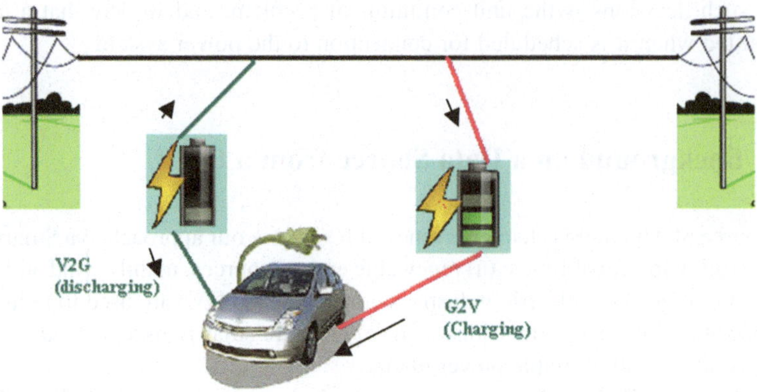

Fig. 8.2 V2G and G2V modes

commitment data set from the Gridable Vehicles defined in [SVY10] and also the NYISO data set [MEG11, NYI].

The rest of the chapter is sectioned as follows: Objectives, Decision-Tree Methods, Introduction to the Weka Tool, Results, and Conclusion.

8.3 Objectives

We have multi-fold goals in this chapter. The first one is to discover how decision-tree models, such as the M5 model, are applicable in power-domain specifically problems such as unit-commitment and how in NYISO data set where demand level can be forecasted based on the history of logs of committed demand data [NYI]. Secondly, we have shown how decision-tree models can help the decision support for system operators who need to do switching operations or to activate or deactivate circuit breakers by exploring the data obtained with the Gridable Vehicles or other data sources. We have also presented the predictive and classification accuracy of the M5 tree-based model in the power-system domain when compared to the Relative Absolute Error (RAE) and Root Mean Squared value (RMSE) metrics of the J48 and decision-stump approaches.

8.4 Classifying and Predicting the Missing Values Using Decision-Tree Models

Data mining studies the algorithms and computational paradigms that allow computers to discover the databases' structure; to perform prediction, classification, and forecasting; and, generally, to improve performance through interaction with the data. Machine learning is concerned with building computer systems that have the ability to improve their performance in a given domain through experience. In this chapter, we use M5P (where the "P" stands for "prime") to generate M5 model trees using the M5' algorithm which was introduced by Wang and Witten (1997) and enhances the Quinlan's [Qui92] original M5 algorithm. We also show the applicability of the J48 and Decision-Stump methods. The Decision-Stump approach builds simple, binary decision "stumps" (one-level decision trees) for both numeric and nominal classification problems. It copes with missing values by extending a third branch from the stump, in other words treating the "missing" as a separate attribute value. Decision Stumps implement trees with a single split only, which are frequently used as base learners for meta learners such as Boosting.

Decision-tree models offer several ways of dealing with missing values that can often minimize or eliminate those values' effect on model performance. The authors believe that the applicability of such decision-tree models in the Smart-Grid platform has not been investigated at system-operator levels, which has tremendous benefits to understand the demand levels from multiple generators, the customers' power use, and other related parameters.

8.5 Model Decision Trees

8.5.1 M5 Model Trees

The M5 model-tree algorithm was originally developed by Quinlan [Qui92]; we used the software and implemented its M5 variation that was provided by Witten and Frank (2000). Model trees combine a conventional decision tree with the possibility of generating linear-regression functions at the leaves. This representation is relatively perspicuous because the decision structure is clear and because regression functions do not normally involve many variables. The M5 tree is a piecewise linear model, so it takes an intermediate position between the linear models and the truly nonlinear models as ANNs.

The construction of a model tree is similar to that of a decision tree. Figure 8.3 illustrates how the splitting of space is done. First, the initial tree is built, and then, the initial tree is pruned (reduced) to overcome the over-fitting problem (when a model is very accurate with the training data set and fails with the test set). Finally, the smoothing process is employed to compensate for the sharp discontinuities between adjacent linear models at the leaves of the pruned tree. (This operation is not needed to build the decision tree.)

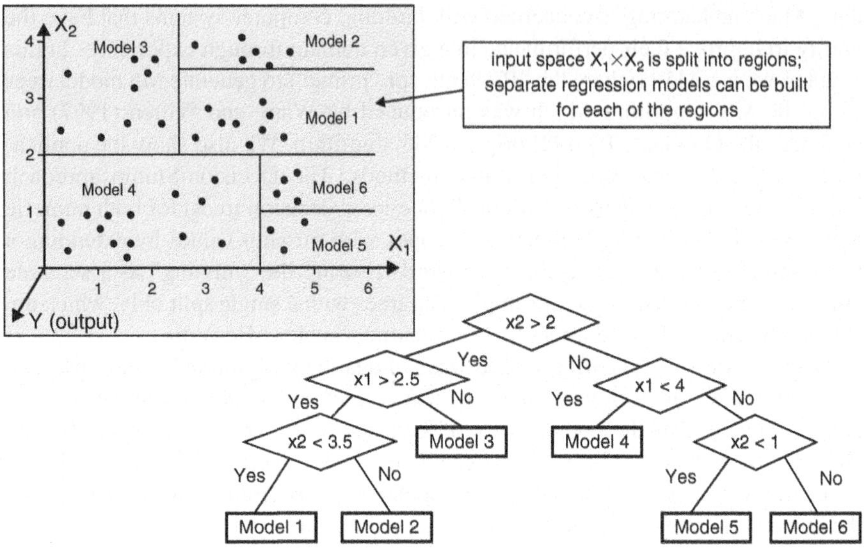

Fig. 8.3 Example of an M5 model tree: Models 1–6 are linear-regression models

8.5.2 *Related Work on Building Decision-Tree Models*

Data-driven modeling based on the advances of machine learning and computational intelligence proved to be a powerful approach for a number of problems [Sol02]. One of the most frequently and successfully used techniques in this respect is an artificial neural network (ANN). It has been demonstrated that there is an entire set of other methods that can be at least as accurate and have additional advantages [SD03]. One such numerical prediction (regression) method that we found to be practically unknown to practitioners is Quinlan's so-called M5 model tree [Qui92]. It is based on ideas from a popular classification method, a decision tree that follows the principle of the input space's recursive partitioning by using entropy-based measures and assigning class labels to the resulting subsets.

Various decision-tree inductive algorithms have emerged in the past decades; these algorithms are used to solve classification problems that primarily employ the divide-and-conquer approach [BF96]. First, an attribute is selected and placed at the root node, and one branch is made for each possible value; then, the example set is split into subsets, with one for every value of the attribute. Now, the process can be repeated recursively for each branch by only using those samples that actually reach the branch. If, at any time, all samples at a node have the same classification, the development of that part of the tree is stopped. We ran these algorithms in WEKA because it has a collection of data-mining methods that are already available.

Viewer

Relation: gv_data_ieee

No.	Time Numeric	U-1 Numeric	U-2 Nominal	U-3 Numeric	U-4 Numeric	U-5 Numeric	U-6 Numeric	U-7 Numeric	U-8 Numeric	U-9 Numeric	U-10 Numeric	V2G/G2V Numeric	Solar Numeric	Wind Numeric	Capacity Numeric	Demand Numeric	Reserve Numeric	Emission Numeric
1	1.0	455.0	154.2	0.0	130.0	0.0	0.0	0.0	0.0	0.0	0.0	-50.01	0.0	10.73	1090.0	700.0	390.0	655.056
2	2.0	455.0	156.8	0.0	130.0	0.0	0.0	0.0	0.0	0.0	0.0	-14.32	0.0	22.45	1054.3	750.0	304.3	656.939
3	3.0	455.0	150.0	114.1	130.0	0.0	0.0	0.0	0.0	0.0	0.0	-24.09	0.0	25.05	1194.1	850.0	344.1	701.944
4	4.0	455.0	232.9	130.0	130.0	0.0	0.0	0.0	0.0	0.0	0.0	-23.07	0.0	25.1	1193.1	950.0	243.1	794.0416
5	5.0	455.0	255.7	130.0	130.0	25.0	0.0	0.0	0.0	0.0	0.0	-21.14	0.0	25.35	1353.1	1000.0	353.1	847.858
6	6.0	455.0	352.7	130.0	130.0	25.0	0.0	0.0	0.0	0.0	0.0	-18.28	0.0	25.5	1350.3	1100.0	250.3	1008.261
7	7.0	455.0	398.0	130.0	130.0	25.0	0.0	0.0	0.0	0.0	0.0	-13.59	0.0	25.5	1345.6	1150.0	195.6	1103.328
8	8.0	455.0	388.8	130.0	130.0	25.0	0.0	0.0	0.0	0.0	0.0	28.48	17.14	25.5	1360.5	1200.0	160.5	1082.973
9	9.0	455.0	431.9	130.0	130.0	25.0	20.0	25.0	0.0	0.0	0.0	26.48	31.5	25.14	1523.5	1300.0	223.5	1234.73
10	10.0	455.0	455.0	130.0	130.0	87.7	20.0	25.0	10.0	0.0	0.0	25.63	36.18	25.41	1577.6	1400.0	177.6	1322.868
13	13.0	455.0	455.0	130.0	130.0	98.2	20.0	25.0	10.0	0.0	0.0	14.78	36.85	25.13	1566.8	1400.0	166.8	1325.569
14	14.0	455.0	429.2	130.0	130.0	25.0	20.0	25.0	0.0	0.0	0.0	28.93	31.84	25.07	1525.9	1300.0	225.9	1228.074
15	15.0	455.0	419.5	130.0	130.0	25.0	0.0	0.0	0.0	0.0	0.0	10.34	9.72	20.42	1342.3	1200.0	142.3	1152.787
16	16.0	455.0	298.5	130.0	130.0	25.0	0.0	0.0	0.0	0.0	0.0	-16.3	12.87	14.87	1348.3	1050.0	298.3	911.372
17	17.0	455.0	246.0	130.0	130.0	25.0	0.0	0.0	0.0	0.0	0.0	-11.09	0.0	25.01	1343.1	1000.0	343.1	835.018
18	18.0	455.0	359.4	130.0	130.0	25.0	0.0	0.0	0.0	0.0	0.0	-18.77	0.0	19.29	1350.8	1100.0	250.8	1021.514
19	19.0	455.0	362.1	130.0	130.0	25.0	20.0	25.0	0.0	0.0	0.0	27.55	0.0	25.41	1524.5	1200.0	324.5	1078.784
21	21.0	455.0	446.8	130.0	130.0	25.0	20.0	25.0	0.0	0.0	0.0	42.74	0.0	25.5	1539.7	1300.0	239.7	1271.883
22	22.0	455.0	359.0	130.0	130.0	25.0	0.0	0.0	0.0	0.0	0.0	-20.32	0.0	21.24	1352.3	1100.0	252.3	1020.729
23	23.0	455.0	227.6	130.0	130.0	0.0	0.0	0.0	0.0	0.0	0.0	-42.64	0.0	0.0	1212.6	900.0	312.6	787.681
24	24.0	455.0	150.0	110.0	127.9	0.0	0.0	0.0	0.0	0.0	0.0	-45.13	0.0	2.26	1215.1	800.0	415.1	697.056
11	11.0	455.0	455.0	130.0	130.0	124.5	20.0	25.0	10.0	10.0	0.0	27.08	38.31	25.01	1634.1	1450.0	184.1	1367.224
12	12.0	455.0	455.0	130.0	130.0	148.3	20.0	25.0	10.0	10.0	10.0	46.41	35.39	24.88	1708.4	1500.0	208.4	1412.925
20	20.0	455.0	455.0	130.0	130.0	116.3	20.0	25.0	0.0	10.0	0.0	40.33	0.0	18.27	1592.3	1400.0	192.3	1333.992

Fig. 8.4 Unit commitment data set

8.5.3 *WEKA Platform*

Weka is a collection of machine-learning algorithms for data-mining tasks. The algorithms can either be applied directly to a dataset or called from your own Java code. Weka contains tools for data pre-processing, classification, regression, clustering, association rules, and visualization. It is also well-suited for developing new machine-learning schemes [MEG11]. Figure 8.4 shows the data set for the unit commitment problem.

Figures 8.3 and 8.4 show the classification of the GV dataset's decision trees into five linear models: LM1, LM2, LM3, LM4, and LM5. For example, in Fig. 8.6, there are ten values in the wind data set that exceed 12.8 MW of power and three values that fall below 12.8 MW, which is useful information for power-system operators to identify the demand scarce, redirection of power, activating and deactivating the generators, and loads. Figure 8.4 shows the unit commitment data from [SVY10].

8.5.4 *Data Pre-Processing in Weka*

Weka's pre-processing capability is encapsulated in an extensive set of routines, called filters, that enable data to be processed at the instance- and attribute-value levels. The most important filter algorithms that are included in WEKA are as follows:

weka.filter.AddFilter
weka.filter.DeleteFilter
weka.filter.MakeIndicatorFilter

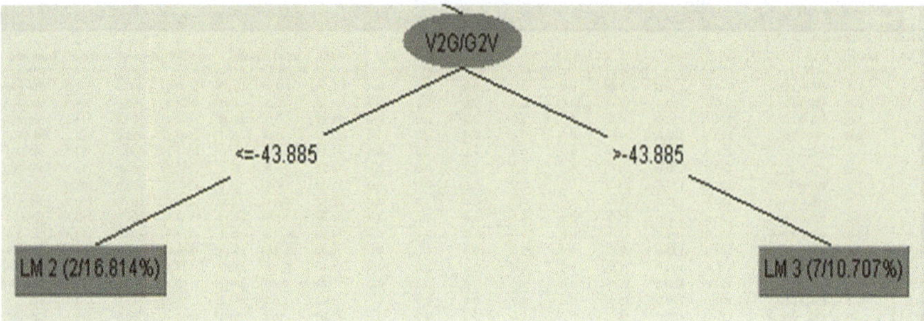

Fig. 8.5 An LM decision-tree branch on V2G/G2V data

weka.filter.MergeAttrbuteValuesFilter
weka.filter.NominalToBinaryFilter
weka.filter.SelectFilter
weka.filter.ReplaceMissingValuesFilter
weka.filter.SwapAttributeValuesFilter
weka.filter.DiscretiseFilter
weka.filter.NumericTransformFilter

Building applications with WEKA is easier. In most data-mining applications, the machine-learning component is just a small part of a far-larger software system. To accommodate this, it is possible to access Weka's programs from inside one's own code. This allows the machine-learning sub-problem to be solved with minimal additional programming.

Figures 8.5 and 8.6 illustrate how the data set can be broken into sub-linear models with a stopping criterion. In Fig. 8.5, there are seven instances where attribute values showing the discharging process occur and where the power is extracted from the electric grid, and there are two instances where charging occurs, meaning that the power is inserted back to the electric grid, thereby acting as a mobile, renewable storage unit. Such instances would help decision makers (system operators) to understand their microgrid environments.

The attribute, which is chosen to be used for a split for a given set of samples, can be determined by the splitting criterion. For decision trees, the splitting is based on trying to minimize the entropy in the resulting subsets, in other words, trying to filter as many samples as possible from the same class into one subset.

In other words, the M5 model tree is a numerical prediction algorithm, and its splitting criterion is based on the standard deviation of the values in subset T of the training data that reach a particular node (which is an analog of entropy). It is used as a measure of the error at that node, and the attribute that maximizes the expected error reduction is chosen for splitting at the node. Accordingly, in Fig. 8.6, the *Reserve* attribute is selected for the root node with the split value 11.5. The splitting process terminates when the output values for the samples that reach a node vary slightly, that is, when their standard deviation is just a small fraction (less than 5%) of the standard deviation for the original sample set. Splitting also terminates when

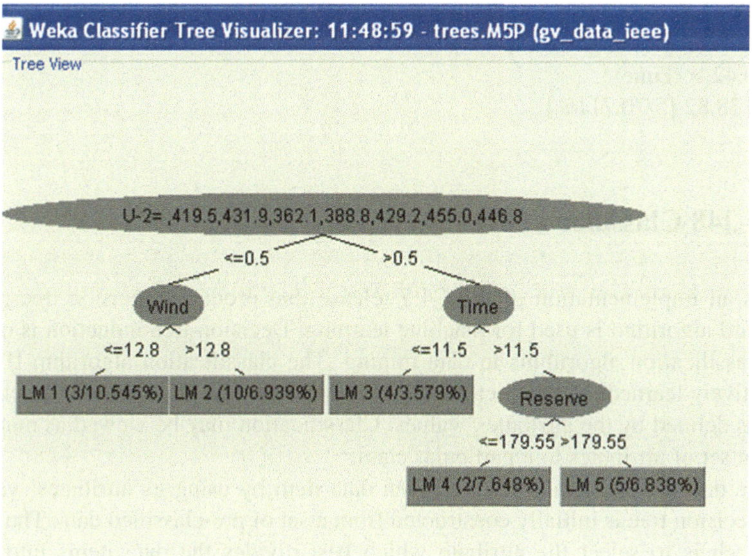

Fig. 8.6 An LM decision tree with reserve, time, and wind data

only a few samples remain in a subset. The linear-regression models are then built for each subset of samples that are associated with the terminating (leaf) nodes.

8.5.5 Pruning and Smoothing the Model Trees

If a generated tree has too many leaves, it may be "too accurate" and, hence, overfit; therefore, it is a poor generalizer. It is possible to make a tree more robust by simplifying it, i.e., by pruning to merge some of the lower subtrees into one node. The process used to compensate for the sharp discontinuities that will inevitably occur at the leaves of the pruned trees between adjacent linear models is commonly referred to as the smoothing phenomenon. Experiments show that smoothing substantially increases the accuracy of prediction. For example, M5 pruned model rules to the "Time" attribute (using smoothed, linear models) results in two rules as follows:

Number of Rules: 2.
Rule: 1
IF
Time <= 19.5
THEN
U-4 =
−0.0093 × Time
+ 130.0772 [19/0%]

Rule: 2
U-4 =
−0.42 × Time
+ 138.82 [5/70.711%]

8.6 J48 Classifier in Weka

J48 is an implementation of the C4.5 release that produces decision trees. This standard algorithm is used for machine learning. Decision-tree induction is one of the classification algorithms in data mining. The classification algorithm [PR] is inductively learned to construct a model from the pre-classified data set. Each data item is defined by the attributes' values. Classification may be viewed as mapping from a set of attributes to a particular class.

The decision tree classifies the given data item by using its attributes' values. The decision tree is initially constructed from a set of pre-classified data. The main approach is to select the attribute which best divides the data items into their classes. According to the attributes' values, the data items are partitioned. This process is recursively applied to each partitioned subset. The process terminates when all the data items in the current subset belong to the same class. Table 8.1 shows a data set taken from the NYISO website.

Figure 8.7 shows a graphical view of the 24-h demand data for New York City in Mega Watts. Table 8.2 shows some specific instances (12 a.m., 1 a.m., and 2 a.m.) of the time attribute for the NYISO data in the WEKA explorer. We used these instances to apply and test the J48, M5, and decision-stump methods. Figure 8.8 is a decision tree that is run using the J48 classifier on WEKA. Each tree branch has a condition as follows: 200 MW<demand<660 MW of power (*attribute* $\leq$ *value* or *attribute* > *value*).

8.6.1 Situational Awareness (SA)

PSOs can better understand situational awareness by mining these sensor data, arriving at good estimate and control mechanisms to take the necessary switching actions. Despite technological upgrades and advances, power-system operators must assimilate overwhelming amounts of data in order to keep the electric utility grid operating. Studies of recent blackouts have demonstrated the need to enhance the operator's ability to understand the system's state and to anticipate possible problems. With the grid's increasing complexity and interconnectivity, the scope and complexity of power-grid operations continue to grow. To confront this escalation, new paradigms are needed to guide the research, tool development, and training in order to enhance and improve operations. This study applies classification theories for decision making to a situational-awareness

Table. 8.1 NYISO data set from April 27, to May 3, 2011, for NYC

Time stamp	27-Apr	28-Apr	29-Apr	30-Apr	1-May	2-May	3-May	Average
0:00	4808	4789	4772	4620	4572	4598	4842	4714.429
1:00	4568	4585	4525	4408	4296	4329	4461	4453.143
2:00	4409	4402	4323	4256	4123	4145	4247	4272.143
3:00	4320	4302	4207	4163	4017	4059	4142	4172.857
4:00	4303	4284	4178	4129	3971	4060	4126	4150.143
5:00	4461	4442	4312	4150	3975	4237	4291	4266.357
6:00	4890	4860	4673	4213	3973	4638	4716	4566.143
7:00	5504	5457	5210	4392	4082	5260	5850	5036.429
8:00	6050	5988	5688	4654	4260	5763	5355	5465.429
9:00	6426	6349	6018	4934	4499	6117	6210	5793.286
10:00	6614	6533	6187	5140	4707	6302	6396	5982.714
11:00	6702	6615	6266	5249	4353	6373	6503	6031.571
12:00	6724	6640	6285	5279	4943	6409	6588	6124
13:00	6733	6651	6304	5271	4989	6436	6672	6150.857
14:00	6720	6644	6294	5234	5002	6446	6722	6151.714
15:00	6698	6634	6267	5196	5003	6448	6745	6141.571
16:00	6675	6621	6237	5162	5011	6466	6776	6135.429
17:00	6633	6571	6172	5142	5058	6482	6743	6114.429
18: 00	6408	6328	5947	5118	5088	6337	6485	5958.714
19:00	6252	6169	5303	5156	5173	6241	6314	5374
20:00	6241	6152	5807	5289	5314	6219	6234	5893.714
21:00	6042	5958	5651	5234	5323	6092	6084	5769.143
22:00	5718	5642	5389	5068	5174	5788	5768	5506.714
23:00	5247	5192	5040	4821	4901	5351	5339	5127.286

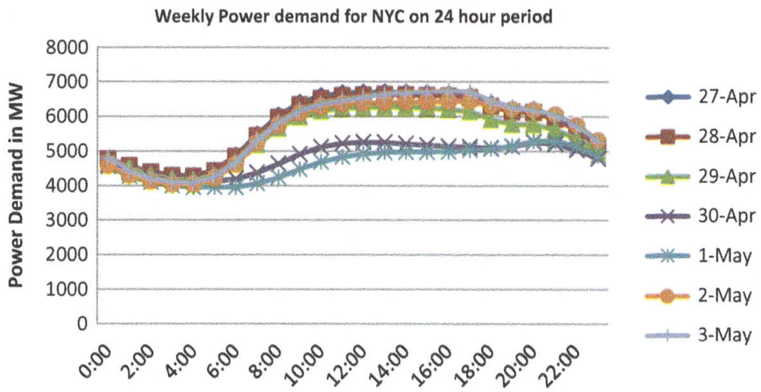

Fig. 8.7 Graph for the NYISO data set

Table 8.2 A sample data set showing 12 a.m. (t0), 1 a.m. (t1), and 2 a.m. (t2)

Time (s)	Demand (in Mega Watts of power)	Date
t0	4808	ap 27
t0	4789	ap 28
t0	4772	ap 29
t0	4620	ap 30
t0	4572	1-May
t0	4592	2-May
t0	4714	Average
t1	4568	ap 27
t1	4585	ap 28
t1	4525	ap 29
t1	4408	ap 30
t1	4296	1-May
t1	4329	2-May
t1	4469	Average
t2	4409	ap 27
t2	4402	ap 28
t2	4323	ap 29
t2	4256	ap 30
t2	4123	1-May
t2	4145	2-May
t2	4247	Average

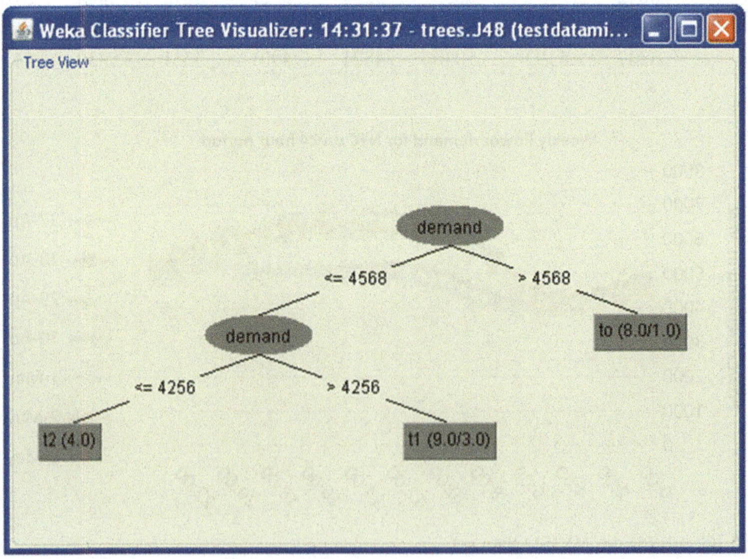

Fig. 8.8 J48 on the NYISO data

(SA) problem from a power-grid perspective and offers a framework to guide the development of tools to increase the grid operator's SA and to enhance operational performance [FGR09].

A sample run of the J48 classifier with the NYISO data is shown below:

```
=== Run information ===

Scheme:       weka.classifiers.trees.J48 -C 0.25 -M 2
Relation:     testdatamining
Instances:    21
Attributes:   3
              time
              demand
              date
Test mode:    8-fold cross-validation

=== Classifier model (full training set) ===

J48 pruned tree
------------------

demand <= 4568
|   demand <= 4256: t2 (4.0)
|   demand > 4256: tl (9.0/3.0)
demand > 4568: to (8.0/1.0)

Number of Leaves  :     3

Size of the tree :      5
```

Here, new notations are used for convenience. For example, "april a" refers to April 27, "april b" is April 28, "april c" refers to April 29, and "april d" is April 30.

8.7 Results and Discussion

The M5 models provide some advantages from both quantitative and qualitative points of view when compared with simpler regression methods. From a quantitative point of view, we can conclude that the goodness, i.e., estimation accuracy, of M5 is better than the classical linear regression or least mean squares (Table 8.3) as seen through Relative Absolute Error (RAE) measures. Various research studies have suggested that selecting an error measure has an important effect on the

Table 8.3 Classification accuracy metrics

MMethod	RMSE	Correlation coefficient	Relative absolute error	Data set
J48	.3037	–	51.3%	NYISO
M5	54.4	.968	14.5%	Unit commitment
Decision stump	125.19	.816	62.79%	Unit commitment
M5	17.6	.99	8.68	NYISO
Decision stump	.3407	–	64%	NYISO

conclusions about which set of classification or forecasting methods is most accurate [Arm85, AC92, Fil92]. They conclude that the Relative Absolute Error (RAE) is a useful measure, especially when making comparisons across a small set of time-series data. The M5 algorithm greatly improves the estimates to levels which are acceptable for the engineering community. Also, when compared with other techniques, the M5 algorithm handles both continuous and categorical variables.

In addition to greater accuracy, the M5 model provides other qualitative advantages. First, each subset is clearly defined in the sense that new instances are easily assigned to a local model. Second, decision trees are easily understood by users, in general, and by project managers, in particular, because we can read them as rules. Such conditions in the power system can provide a clear indication about which variables are most important for prediction (as conditionals in the tree branches). The tree's leaves allow the grid's system operators to gain further knowledge about the dataset's characteristics. The following shows the classification results based on the demand using the M5 and decision-stump methods. It is evident that the decision-stump and J48 methods yield marginal classification and a high relative absolute error (RAE).

8.8 Conclusion

We showed how the collected Smart-Grid data can be analyzed with the M5 algorithm for easier interpretation for system operators to build decision-support engines. This chapter presented the results of applying the M5 algorithm as an estimation technique to two weekly datasets generated from the power system's repository. We compared the classification results for the J48, M5, and stump methods and found that there is improved accuracy with the results when using the M5 model. Furthermore, the M5 method can help understand the datasets better; the tree (rules) generated with the M5 method provides system operators with a better understanding about what attributes are more important for a particular dataset that can help influence a decision. Finally, we believe that the applicability of the M5 method, in particular, to the power-system domain, in general, provides certain advantages for power-system operators while keeping the estimation process within reasonable complexity.

Chapter 9
Maximization of the Utility Function, Time-Dependent Energy Allocation, and Fuzzy-Logic Resource-Allocation Models

9.1 Introduction

Dynamic, real-time power systems often operate in continuously changing environments such as adverse weather conditions, sudden transformer failures, or malfunctioning of a sub-system of a transmission or distribution network. These disruptions, along with the power-network systems' complexity, cause the system's energy demand and loads to fluctuate, potentially resulting in widespread outages and huge price spikes. Electric energy is traditionally generated at large power plants and transferred, in bulk, to substations that distribute power to residential, commercial, and industrial customers. A wide variety of devices are involved and utilized, including transformers; sensors; controllers; and, more recently, smart meters at the consumption points. A successful Smart Grid must be highly instrumented, data rich, networked, and integrated and managed as a complete "end-to-end" system. However, centralized control is clearly impossible, leading to the challenge of distributing control while still providing high performance and efficiency.

Data from the North American Electric Reliability Council (NERC) and analyses from the Electric Power Research Institute (EPRI) indicate that average outages from 1984 to the present time have affected nearly 700,000 customers per event annually as noted by Masood Amin [Ami04]. Smaller outages occur much more frequently and affect tens to hundreds of thousands of customers every few weeks or months while larger outages occur every 2–9 years and affect millions. Although preventing these outages remains challenging, the consumers' demand changes (increases or decreases) can often be offset by distributed energy resources (DERs), renewable resources such as solar and wind-based power, in order to satisfy the shortages or to reduce the outage levels. In our work, we consider using such DER-based standby mechanisms and formulate to support their issue.

P. Ranganathan, K.E. Nygard, *Distributed Linear Programming Models in a Smart Grid*, Power Electronics and Power Systems, DOI 10.1007/978-3-319-52617-1_9

9.2 Background

Green-energy solutions that are described as "smart"—smart meters, smart buildings, smart appliances, and the Smart Grid—have intelligent sensors to measure temperature or other variables to receive and transmit data, memory chips to store the information and process it, and manage the power flow to adjust energy loads. When building a Smart-Grid self-healing model, there are multiple considerations that are important to include as explained by Amin. Some considerations pertain to the physical infrastructure, such as the generators, busses, relays, and transmission lines. Other considerations pertain to the cyber infrastructure, such as the communication networks, storage, protocols, and procedures for grid management. There have been several studies that push for advanced, secure grid infrastructure as described by Amin and Stringer.

The current electric grid in the U.S. is about 100 years old, and while incremental upgrades have been made over the last few decades, advancements in fields such as software and internet technologies, and wireless technologies, have been slow to transition into the grid. Today, the convergence of several factors provides for a perfect storm that is expected to make a revolutionary change in the nature of the electrical grid.

The various factors that led to the futuristic, self-healing grid are as follows:

- Rapid reduction in limited natural resources
- Population growth
- Rising cost of energy
- Current technology advances make it possible
- Increased global warming and climatic conditions
- Changes in the utility-business operations worldwide
- Deregulation in parts of the world
- Better awareness and education among consumers
- Increased renewable, distributed, and smaller power generation
- Increased power-storage capability
- Radio frequency IDs (RFIDs) and sensors that have narrowed the virtual distance between the physical and cyber worlds

As a result, there is tremendous pressure at the government, industry, and consumer level to innovate. Various entities are working and have articulated their visions about what the futuristic grid should look like.

Today's grid features a typical, centralized approach where few powerful central stations broadcast energy to different consumers. As renewable energy resources pave their way, we expect that future users will not only be energy consumers, but also producers; they reflect the ability to interchangeably slip between these roles. The smart grid defined in Amin is an emerging concept with the goal to provide the next-generation electricity network that will boast advanced configurability, reactiveness, and self-manageability. It is a complex infrastructure that depicts a system of system characteristics, such as the interdisciplinary nature, the elements' operational and managerial independence, geographical distribution, high heterogeneity of the networked systems, and emergent behavior and evolutionary development. It is expected to be the key part in a global ecosystem of interacting entities where cooperation will give birth to innovative cross-industry services. Key driving forces behind these efforts are energy efficiency and better management of the available resources (locally and globally). In order to achieve this goal, fine-grained monitoring and management are needed. Primary considerations for a self-healing, cyber-grid architecture are given in the following sections.

Identification of malfunctions. When an electricity supply/demand problem occurs, it must be accurately and rapidly identified in terms of its location, type, and severity. Smart meters that are programmed to report energy consumption at regular intervals help to pinpoint problems by simply reporting that, suddenly, no electricity is being consumed at specific sites. Phasor Measurement Units (PMUs) that are located at interconnection points in the grid and that are synchronized in time using the Global Positioning System (GPS) can report phase-angle differences that are associated with unstable conditions. The precise unit timing allows for wide-area sensing and monitoring at rates on the order of 60 samples per second. A Phasor Data Concentrator (PDC) can receive data from multiple PMUs and can then correlate, store, and trigger response actions to abnormal conditions. Networks of smart meters and PMUs, along with other sensing and reporting devices, make it possible to quickly and accurately determine the details of the grid's problems.

Data storage. At least 33 U. S. states had smart meters installed in 2010, and an estimate by the Edison Electric Institute indicates that 60 million will be installed by 2019. The large volume of data from the grid-sensing devices underscores the need for large-scale, distributed database systems in the Smart Grid. The nodes that store data may be capable of carrying out the data mining to predict future problems, to send alerts, and to make resource-allocation decisions autonomously.

Energy storage, by itself and in combination with distributed generation (termed ES-DER), is a new and emerging technology that has been identified by FERC as a key functionality of the smart grid, and standards related to storage should be treated as a key priority by the Institute and industry in the interoperability standards development process, subject to certain reservations. Coupled with inverter-based technology, these systems can be used to improve EPS performance. Due to the infancy of using storage and inverter technologies as a grid-integrated operational asset, there are few standards to capture how it could or should be utilized on the legacy grid and Smart Grid. For example, to date, there is no guidance or standards to address grid-specific aspects of aggregating large or small mobile storage, such as Plug-in Hybrid Electric Vehicles (PHEVs). ES-DER is treated as a distributed energy resource in some standards, but there may be distinctions between electric storage and connected generation. In particular, storage-based systems may function as a load more than 50% of the time.

At the same time, we are moving toward a large penetration of renewables in the grid, which could be destabilizing, but should, in the context of the Smart Grid, allow these renewables to be true utility assets. The potential for instability is twofold: first, due to the intermittent nature of renewables and, therefore, their unsuitability to be dispatchable resources and, second, due to the interconnection regulations that can lead the electronic interconnection interface (the inverter) to trip in response to minor variations with grid voltage or frequency. Because a low frequency is the result of insufficient generation, tripping high-level inverter-based systems would contribute to the problem and cause possible stability issues in response to a relatively minor disturbance. Appropriate interconnection standards, smart-grid devices, and storage are key elements of the solution. Storage allocation can be done using the linear-programming approach.

Utility function. Utility is an economic concept that is used to express the most basic element of well-being for a representative consumer. Utility functions are generally used to show logical inferences about the tradeoffs that consumers make and their likely outcome. The concept of using utility functions to represent consumer *choice* as reflecting likely consumer *behavior* has long been attacked, partially because it appears (wrongly) to assume that economic actors are rational and partially because it is tautological.

A utility is a numerical rating that is assigned to every possible outcome which a decision maker may face. In a choice among several alternative prospects, the one with the highest utility is always preferred. To qualify as a true *utility* scale, however, the rating must be such that the utility of any uncertain prospect is equal to the expected value (the mathematical expectation) for the utilities of all its possible outcomes which could be either "final" outcomes or uncertain prospects. When decisions are made by a so-called *rational agent* (If A is preferred to B and B to C, then A must be preferred to C.), it should be clear that *some* numerical scale can be devised to rate any possible outcome "simply" by comparing and ranking these. Determining equivalence in monetary terms may be helpful with such a systematic process, but it is not theoretically indispensable. What may be less clear, however, is how to devise such a rating system so that it possesses the above fundamental property that is required for a *utility* scale.

Measuring the benefit that is obtained with a management action is the well-studied role of the utility function. A common convention is to normalize the utility function into the range of 0–1, with the 0 end of range modeling the minimum benefit.

Application conflicts. We presume that distributed software agents will run applications at various nodes in the grid in order to summarize and provide information, to mine data, and to invoke certain control actions. Some applications' polling actions will compete to acquire the necessary data and bandwidth, degrading each other's performance. To maintain Quality of Service (QoS) standards, the applications will require resource allocations.

9.2.1 Large-Scale Resource Optimization

There are significant needs and benefits to address (1) large-scale supply chain optimization problems; (2) product pipeline-management (Here, we refer to them as '*agents.*') decision making; (3) the detailed optimization of process models; (4) the design and optimization of the network's nodes; and (5) optimization of micro-grid models, such as small-scale or RTU designs, at the strategic, tactical, and operational levels. These static and dynamic optimization models, which commonly have hundreds of thousands of variables, include non-linear as well as mixed-integer programming problems. These models can be posed in equation form or with black-box simulation software. Rigorously handling non-convexities for the nonlinear models has emerged as a major challenge that will involve global-optimization methods that are potentially very expensive.

9.3 Related Research

Mathematical programming has enjoyed a burgeoning presence in theoretical computer science, both as a framework for developing algorithms and, increasingly, as a bona fide computation model. The limits are expressed in terms of sizes of formulations and integrality gaps of formulations as the one described in Arora. Linear formulations are an appealing computation model because both optimization and decision problems fit naturally into the framework, and both theoretically tractable and efficient practical algorithms exist to solve the linear program as defined by R. E. Moore [Moo91].

Many practical problems in operations research can be expressed as linear-programming problems. Certain linear-programming cases, such as *network-flow* problems and *multi-commodity-flow* problems, are considered important enough to have generated much research on specialized algorithms for their solution. A number of algorithms for other optimization problems work by solving LP problems as sub-problems. Hanssmann and Hess developed a more comprehensive linear-programming approach for the production-planning problem. This approach became a model for several further research efforts and is the foundation for the simple illustrative model presented in this chapter. A brief overview of this model can be found in McLeavey and Narasimhan (1985).

Although the existence of optimization techniques can possibly be traced to the era of Isac Newton, L. Euler, J. L. Lagrange, and A. L. Cauchy, the development of the simplex method for linear programming by G. B. Dantzig in the mid-1940s, in a sense, started the subject of mathematical optimization the way we understand it today. Another major development was from H. W. Kuhn and A. W. Tucker (1951) who gave necessary and sufficient optimality conditions for nonlinear optimization or non-linear programming problems. Historically, ideas from linear programming have inspired many central concepts of optimization theory, such as *duality, decomposition,* and the importance of *convexity* and its generalizations by Hillar and Gerald (2001). Likewise, linear programming is heavily used in microeconomics and company management, such as planning, production, transportation, technology, and other issues. Although modern management issues are ever-changing, most companies want to maximize profits or minimize costs with limited resources.

Therefore, many issues can be characterized as linear programming. Although, there may be studies using LP for many applications, we believe that no literature has examined grid-resource allocation or the assignment problem using LP in the electric-grid context. This chapter uses LP in the context of a Smart-Grid application.

9.3.1 Collection of Generalized LP Models

Linear programming is a branch of applied mathematics that deals with solving optimization problems of a particular form. Linear-programming problems have a linear *cost* function (consisting of a certain number of variables) which is to be minimized or maximized subject to a certain number of constraints. The constraints are the variables' linear inequalities that are used in the cost function. The cost function is sometimes called the objective function. Linear programming is closely related to linear algebra; the most noticeable difference is that linear programming often utilizes inequalities, rather than equalities, in the problem statement. Linear programming is a considerable field of optimization for several reasons.

Linear programming is a problem-solving technique that involves the optimization of a decision, subject to one or more constraints. The basic logic of the approach is displayed in four steps: (1) recognizing the LP problem, (2) formulating the LP problem, (3) solving the LP problem, and (4) interpreting the LP solution. Several conditions must exist before a problem can be addressed with linear programming. (1) There must be a well-defined single objective that can be stated mathematically as either a minimization or maximization function. (2) There must be a set of decision variables that allow alternative courses of action. (3) There must be constraints for the objective's achievement that are imposed by the availability of scarce resources or other restraints, such as product demand. (4) The objective and all constraints must be able to be expressed as linear functions.

To formulate the LP problem, the example's simple formulation is provided using an objective function and constraints. Depending on the structure of the linear-program formulation, one of several techniques can be applied to solve the LP problem. The most common method is the simplex method, or we can apply branch-and-bound technique that is discussed later in the chapter. To interpret the LP solution, the example's interpretation detail is provided later in the chapter.

We discuss generalizations about the LP models that are applicable to the electric grid (Sect. 9.3.2) as well as the types of problems and constraints which can be handled linearly. Some brief comments about its generalizations to handle situations with multiple constraints are given. The following LP models can be applied to electric-grid loads and supply resources.

9.3.2 *Section 1*

9.3.2.1 Model 1. LP Model by Maximizing Utility

Multi-objective programming involves the recognition that the decision maker is responding to multiple objectives. Generally, objectives are conflicting; not all objectives can simultaneously arrive at their optimal levels. An assumed utility function is used to choose appropriate solutions. Several fundamentally different utility-function forms have been used with multi-objective models. These may be divided into three classes: lexicographic utility, multi-attribute utility, and unknown utility.

The lexicographic utility-function specification assumes that the decision maker has a strictly ordered, pre-emptive preference system among objectives with fixed target levels. For example, a lexicographic system could have its first priority goal as power not less than 10,000 MW, the second priority as leisure that is not less than 20 h a week, the third as a power of no less than 12,000 MW, etc. This formulation is typical of "goal-programming models". The various goals are dealt with in strict sequential order: higher goals before lower-order goals. Once a goal has been dealt with (meeting or failing to meet the target level), its satisfaction remains fixed, and the next lower-order goal is considered. Addressing the lower-level goals does not alter the satisfaction of higher-level goals and cannot damage the higher-level goals with respect to target-level attainment.

Multi-attribute utility approaches allow tradeoffs between objectives to attain maximum utility. The most common form involves maximizing the sum of linearly weighted objectives. This type of formulation has been used by Candler and Boeljhe as well as Barnett, Blake, and McCarl. The third utility approach involves an unknown utility-function assumption. Here, the entire Pareto efficient (non-dominated) solution set is generated so that every solution is reported where one of the multiple objectives is as satisfied as it possibly can be without making another objective worse. We follow Candler's approach for utility functions in this chapter.

We explain LP using utility functions with a simple example as follows. In this linear-programming model that maximizes utility, we consider two basic load types. They are the High-Priority (HP) loads with a hard demand requirement (i.e., Some demand units expect a high QoS that is sensitive to time.) and the Low-Priority (LP) Loads with flexible demand requirements. The load-demand levels, in terms of utility, for these two load types over time are shown in Fig. 9.1.

Assumptions

Here, the model makes the following assumptions:

There are M loads in the electric grid, and it has a high preference for high-priority loads over low-priority loads.

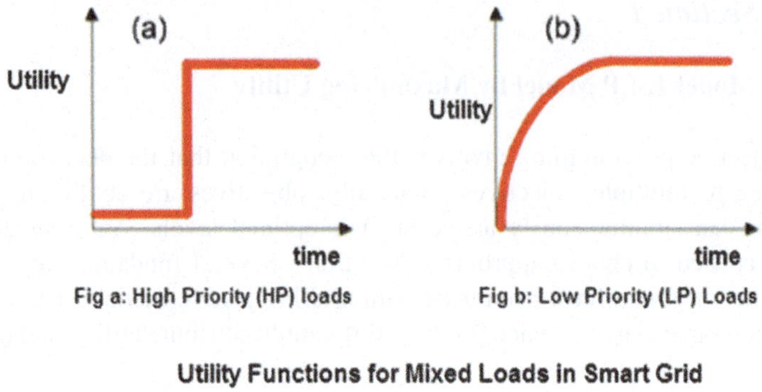

Fig. 9.1 Utility functions for mixed loads in smart grid

Priority: HP loads first, then LP loads.
Pricing: HP loads will pay more.

Two-Step Approach for a Mixed-Demand Load Power Grid

1. Satisfy the Quality of Service (QoS) requirement for HP loads (as many as possible).
   (a) Admission control
   (b) Resource allocation
2. Remaining resources: Distribute them fairly among the LP loads.

$$\max_{Z_{i,k},t_i} \sum_{i\in\{\text{LP loads}\}} V_{\text{LP}}(t_i).$$

Subject to
$t_i \geq Q_i, \quad i \in \{\text{HP loads}\}$ (HP Load Constraints)
Q_i—QoS demand of i - th HP loads

$$t_i = \sum_{l=1}^{L} C_{i,l} Z_{i,l} \; i \in \{\text{loads}\}, l \in \{\text{clusters of mixed loads}\}$$

$C_{i,l}$ —Capacity of the lth cluster on load i

$$\sum_{i=1}^{M\text{loads}} Z_{i,l} = 1, Z_{i,l} \in [0,1]$$

Note: Here, we define clusters by grouping adjacent loads in the grid. One reason for using cluster approximation is that it reduces the number of variables during allocation. Q_i—QoS demand of i-th HP user.

9.3.2.2 Model 2. Time-Dependent Supply-Demand (Energy) Allocation

Each supply source in this resource-allocation model is said to have two parameters: (a) a fixed pricing cost that is associated with employing a selected supply source and (b) a demand meeting time period.

The fixed rental charge represents the cost of renting a resource and is paid once if a resource is used in one or more time periods. The following mathematical formulation characterizes the objective function and the constraints that are required to identify the resource supply that minimizes the sum of all costs. Therefore, we assume that the model has m time periods for which the demand must be met by a selected supply and that there are n demand-fighting resources.

$$\text{Min } z = \sum_{j=1}^{m} \sum_{i=1}^{n} c_i H_j D_{ij}$$

Subject to

$$\forall \sum^{m} D_{i,j} \leq Z_i$$

Decision variables:

D_{ij}—Binary variable that takes on a value of 1 for the time period during which the demand is achieved for the employed resources.

H_j—Time-period counter.

C_i—Cost employed to choose a resource, i.e., a power supply.

Z_i—Binary variable defining whether the i-th resource has been dispatched. Dispatched = 1, not dispatched = 0.

9.3.2.3 Model 3. LP Formulation with the Fuzzy-Rule Selection

Let I be a set of m constrained resources, and let R and $Y \cong (Y_1; Y_2; \ldots, Y_m)$ be $m + 1$ random variables, with R being a reward and Y_i being the amount of resources of type i that are consumed. The joint distribution of R and Y depends on which fuzzy rule is chosen. We make no assumptions about R and Y other than that each one has a known expected value for every fuzzy rule that is selected.

The objective is to choose a fuzzy rule within some finite feasible set S, possibly at random, to maximize $E(R)$ subject to $E(Y) \leq b$, where b is an appropriately dimensioned vector of resources and $E(\ )$ is the expected value operator.

Denote $E(R)$ and $E(Y_i)$ when using rule s as ER_s and EY_{si}, respectively. With this notation, we can express our problem as a linear program, LP(S), where the x_s variable represents the probability of choosing fuzzy rule s:

$$LP(S) : \text{maximize} \sum\nolimits_{s \in S} ER_s x_s$$

subject to

$$\sum_{s \in S} E\,Y_{si} x_s \leq b_i \;\; \forall i \in I$$

$$\sum_{s \in S} x_s = 1, \text{and } x_s \geq 0 \;\; \forall s \in S$$

The sums in the objective function and the resource constraints are $E(R)$ and $E(Y_i)$, respectively, by the conditional expectation theorem. We assume that $b \geq 0$ and that there is a no fuzzy rule exist in S that consumes no resources, so LP(S) has a feasible solution. The value of LP(S) will, in general, be greater than the value of the restricted problem where the variables are required to be 0 or 1; that is, it is significant that the decision maker is permitted to choose a strategy at random as long as the expected resource consumption meets the constraints.

9.3.2.4 Model 4—Uncertainty Handling Through the POMDP Process in an Agent-Oriented Smart Grid

The ability for an agent to reason under uncertainty is crucial for many planning applications because an agent rarely has access to complete, error-free information about its environment. Partially Observable Markov Decision Processes (POMDPs) are a desirable framework in these planning domains because the resulting policies allow the agent to reason about its own uncertainty. In domains with a hidden state and noisy observations, POMDPs optimally trade between actions that increase an agent's knowledge and actions that increase an agent's reward.

Bayesian networks have been widely used for diagnostic purposes [PPM08], [Bou01]. These models can be extended to POMDPs to select the best action in the smart-grid environment. This allows modeling partial observability due to causes and the utility of executing various tests in the electric grid. We describe the problem of refining diagnostic POMDPs when sensor feedback is available. The sensor could be relays, switches, circuit breakers, PMUs, or the system operators who monitor for any grid vulnerability. We propose utilizing sensor feedback to pose constraints on the model, i.e., the transition, observation, and reward functions. These constraints can then be used to efficiently learn the POMDP model and to incorporate expert knowledge about the problem (Fig. 9.2).

The applicability of POMDPs to the renewable-resource allocation problem can be addressed through agent-oriented design. A POMDP involves a sequence of decisions such as activating which relays in the grid, $U_0, \ldots, U_{N-1}$, and a sequence

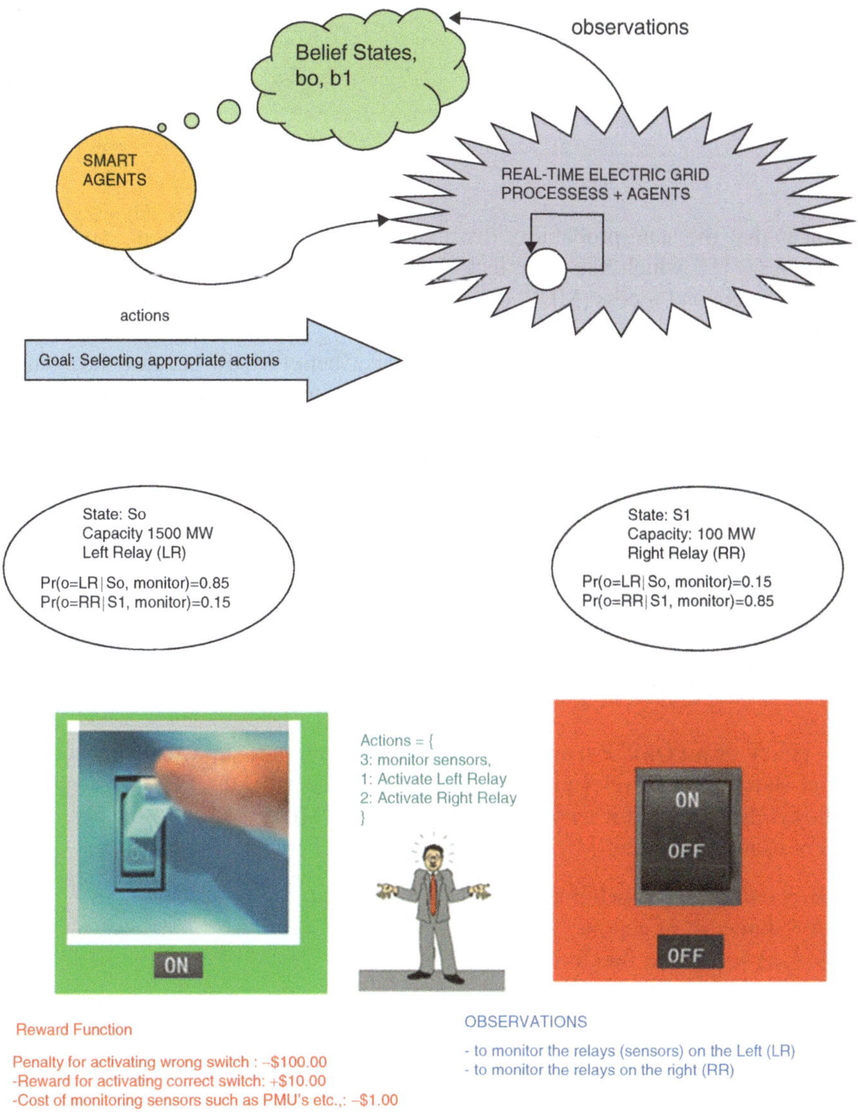

Fig. 9.2 POMDP relay example: an illustration of states, rewards, and observations when switching between higher- and lower-capacity DER sources

of sensor observations, Z_0, Z_2, Z_{N-1}, with a decision-making rule are admissible if U_k depends only on the observable history, $I_k = (Z_0, Z_1, \ldots Z_k; U_0; U_0, \ldots, U_{k-1})$, for $k = 0\ldots N-1$. The Z_k observation depends stochastically on the true state (X_k) of the process at time k, but true state X_k is known only to the extent that it can be deduced from I_k. The set of admissible fuzzy rules, S, in the electric grid is typically enormous, but POMDPs are solvable as a practical matter because of the Markovian nature of state evolution and the way that observations are generated. The crucial

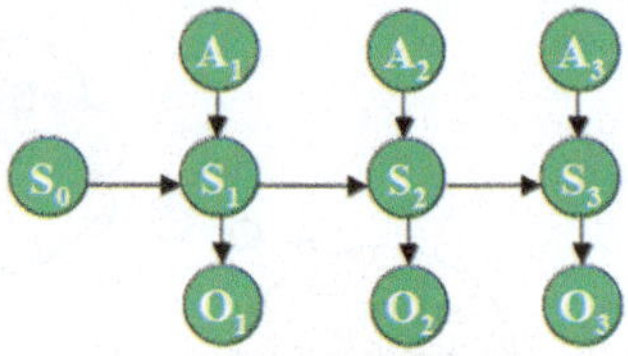

Fig. 9.3 Agents and belief states

result is that the state-probability distribution given I_k is a sufficient statistic for decision U_k [1], which has the effect of converting the POMDP into an ordinary Markov Decision Process (MDP) over a much larger, but observable (via the Bayes Theorem), state space [Bou02].

A POMDP can be seen as a continuous-space "belief MDP" because the agent's belief is encoded through a continuous "belief state." We may solve this belief MDP as before by using the value iteration algorithm to find the optimal policy in a continuous space. However, some algorithm adaptations are needed to fit a resource-allocation problem. There are two regions where uncertainties can exist: uncertainty about the action outcome and uncertainty about the world state due to imperfect (partial) information (Fig. 9.3).

9.3.3 *Belief MDP*

The policy of a POMDP maps the current belief state into an action. Because the belief state holds all relevant information about the past, the POMDP's optimal policy is the solution for a continuous-space belief MDP.

For example, a belief MDP is a tuple <B, A, ρ, P>:

B = infinite set of belief states
A = finite set of actions
$\rho(b, a)$ = (reward function)
$P(b'|b, a)$ = (transition function)
The states' reward and transition function can be defined as follows:

$$\rho(b,a) = \sum_{s \in S} b(s)\, R(s,a) \quad \text{Reward Function}$$

$$\rho(b'|b,a) = \sum_{o \in O} P(b'|b,a,o) P(o|a,b) \quad \text{Transition function}$$

where $P(b'|b, a, o) = 1$ if $\text{SE}(b, a, o) = b'$,
$P(b'|b, a, o) = 0$ otherwise;

9.3.3.1 Transitional Probabilities (Tables 9.1, 9.2 and 9.3)

9.3.3.2 Observational Probabilities (Tables 9.4, 9.5, 9.6 and 9.7)

The belief vectors (b_o, b_1, etc.,) shown in Fig. 9.4 need to be updated at each state based on the transition and observation probabilities, and they should be computed as follows:

Table 9.1 Does not change relay's position

Prob. (monitor)	Relay: left	Relay: right
Relay: left	1.0	0.0
Relay: right	0.0	1.0

Table 9.2 Problem reset

Prob. (left)	Relay: left	Relay: right
Relay: left	0.5	0.5
Relay: right	0.5	0.5

Table 9.3 Problem reset

Prob. (right)	Relay: left	Relay: right
Relay: left	0.5	0.5
Relay: right	0.5	0.5

Table 9.4 Does not change relay's position

Prob. (monitor)	O: LR	O: RR
Relay: left	0.85	0.15
Relay: right	0.15	0.85

Table 9.5 Any observation without monitoring sensors in the grid is uninformative

Prob. (left)	O: LR	O: RR
Relay: left	0.5	0.5
Relay: right	0.5	0.5

Table 9.6 Any observation without monitoring sensors in the grid is uninformative

Prob. (right)	O:LR	O:RR
Relay: left	0.5	0.5
Relay: right	0.5	0.5

Table 9.7 Immediate rewards

Reward (monitoring sensors)		Reward (left relays)		Reward (right relays)	
Relay: left	-1	Relay: left	−100	Relay: left	+10
Relay: right	-1	Relay: right	+10	Relay: right	−100

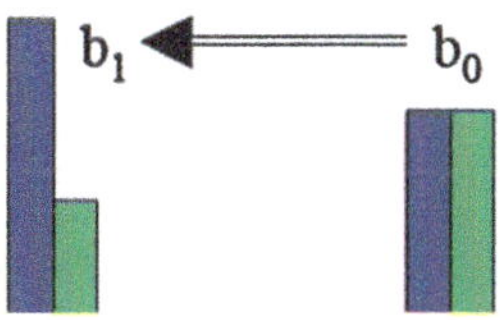

Fig. 9.4 Belief vectors

Belief vectors updated from bo to b1

$$b_1(S_i) = \frac{P(o|S_i, a)\sum_{sj\in S} P(S_i|S_j, a) b_o(S_j)}{P(o|a, b)}$$

The sequence of optimal actions is known as the agent's optimal policy for interacting with its environment. We have two cases to choose an optimal policy for using the POMDP smart-grid example.

Case 1. Optimal Policy at $t = 1$ (Fig. 9.5)

Case 2. Optimal Policy at $t = 2$ (Fig. 9.6)

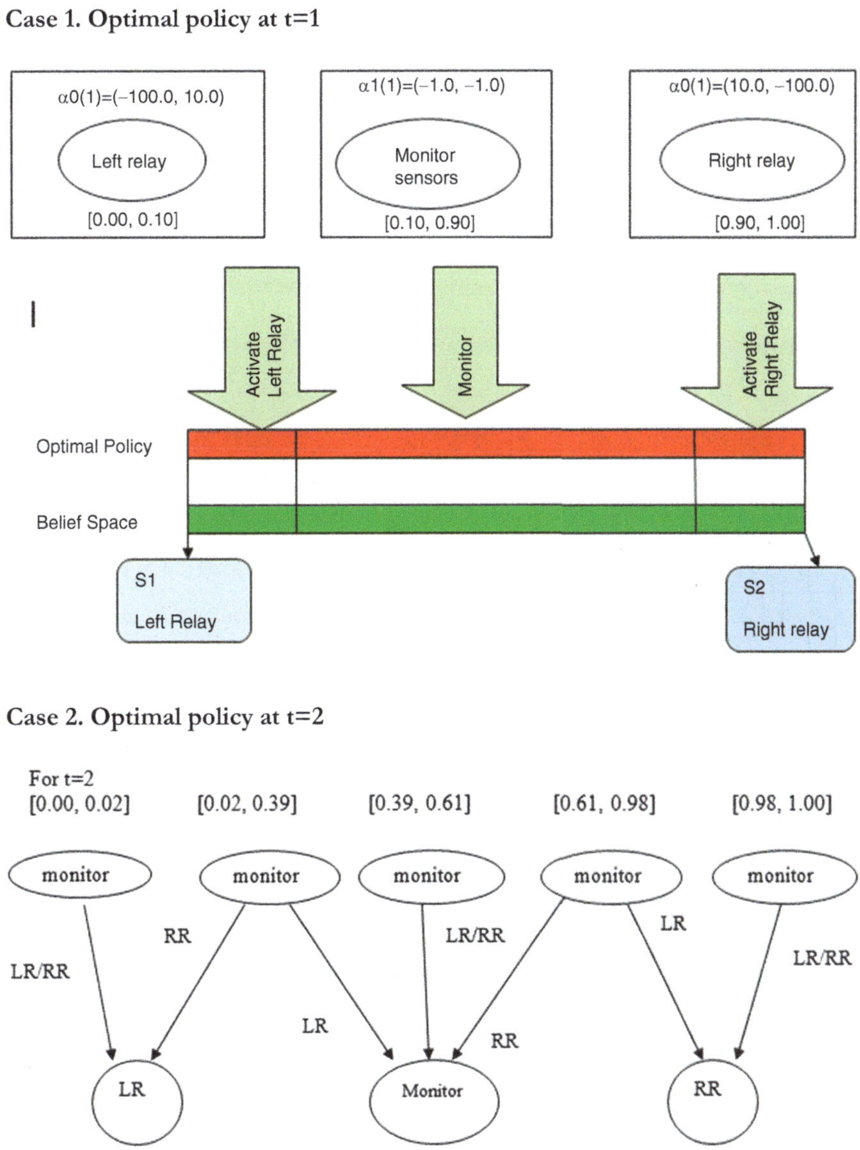

Fig. 9.5 Optimal policy generation flow

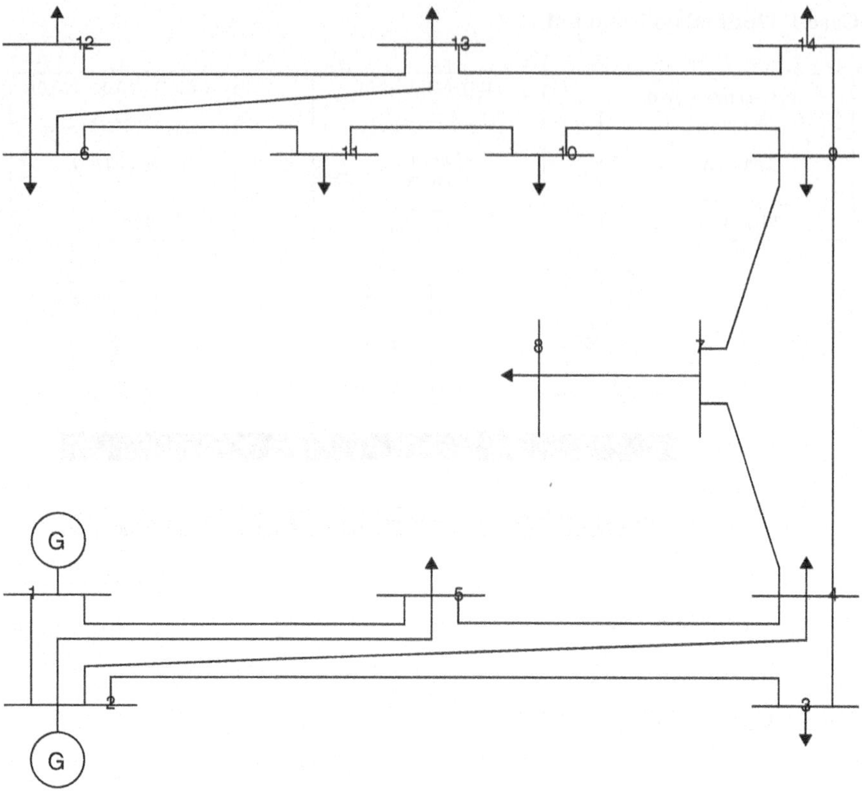

Fig. 9.6 State transitions that show relay switching and agent actions

9.4 Conclusion

In this chapter, we have presented a series of models that are applicable to the resource-allocation problem in a Smart-Grid environment. The presented POMDP approach is efficient in dealing with uncertainties and imperfect information such as on a complex power network application. The feasibility of these approaches needs to be analyzed with real-time, electric-grid data as part of future studies.

Chapter 10
Placement of Synchrophasors Using Linear Programming and Zero-Injection Constraints

10.1 Introduction

Electric-grid parameters, such as the frequency, amplitude, and phase of an electric signal (collectively described by its phasor), must be constantly monitored by special sensors that are usually referred to as Synchrophasors or phasor measurement units (PMUs) [Pha93]. To produce useful data, the PMUs have to be time-synchronized with the Universal Coordinated Time (UTC) with a Global Positioning System's (GPS) receivers or network systems that implement the Precision Time Protocol. One concern with Phasor Measurement Units (PMU) is their location placement. A PMU that is placed on a bus in a test system is capable of measuring currents, voltages, the phasor, and frequencies for an entire transmission line incident to that bus. Using Ohm's law, when a PMU is placed on a bus, the neighboring buses also become observable; i.e., adjacent bus voltages are easily computable. Thus, it is not necessary to place a PMU on every bus in the system. PMUs are expensive sensor units, and depending on the number of measurement channels and the needed features, the cost can rise. The PMU's price is a significant economic constraint for smaller utilities because the communication and IT infrastructure that are required to receive PMU data puts a burden on the total cost of purchasing a PMU unit. Although the usage of PMUs is increasing annually, there is neither a concrete idea nor an effective methodology to place PMUs on the bus systems. Therefore, an appropriate methodology is needed to determine the optimal locations to place these devices for complete observability. The optimal locations can be strategic buses which are very critical to the system. A location is said to be critical if there is large supply of electric generation or demand units.

Various algorithms have been proposed for the placement problems. In [BMB93], a bisecting search algorithm was proposed to find the minimum number of PMUs needed to make the system fully observable. In [NPF05], a Simulated Annealing (SA) technique was proposed in a graph-tree framework to find the minimum number of PMUs. In [MB03], a Genetic Algorithm (GA) was proposed.

P. Ranganathan, K.E. Nygard, *Distributed Linear Programming Models in a Smart Grid*, Power Electronics and Power Systems, DOI 10.1007/978-3-319-52617-1_10

Recent work on optimal PMU placement using an integer linear-programming (ILP) approach was formulated by Abur [Abu04, BYA]. An economic approach for how to schedule PMU placement in multiple stages was proposed [DD08]. An integer quadratic-programming approach was used in [CKE09] for the same OPP problem. The Particle Swarm Optimization (PSO) technique was also used in [HRAS07]. Recently, a chemical-reaction optimization method [WL13] was proposed for the OPP problem.

Literature studies have not considered zero-injection buses and critical buses as constraints, although minimizing the number of PMUs was the objective [SSP12, MH12]. In our model, we have included the following constraints:

1. Zero-injection buses
2. Critical buses

Zero-injection buses are similar to transshipment nodes in an OPP algorithm that reduce the minimum number of PMUs. During contingency situations such as the failure of the PMU, the loss of a communication channel, or the failure of a transmission line, there is a loss of observability. In [MMML13] and [MMK10], the authors consider zero injection as a special constraint in the formulation to keep the system observable during contingency cases. This chapter considers both constraints. The rest of the chapter is organized as follows: Sect. 10.2 introduces the Inter Linear Programming (ILP) formulation for PMU placement problem. In order to validate the proposed algorithm, different test cases are considered in Sect. 10.3. Section 10.3 provides the simulation results and comparative studies with the existing literature. A summary is listed in the Sect. 10.4.

10.2 Placement Problem Formulation

For a system with N buses, the PMU's placement problem can be formulated as follows:

$$\text{minimize} \sum_{i}^{N} w_i \cdot x_i$$

$$\text{s.t. } f(X) \geq \widehat{1},$$

where x_i is a binary decision variable vector where the entries are defined using binary variables as follows:

$$x_i = \begin{cases} 1 & \text{if a } PMU \text{ is installed at bus } i \\ 0 & \text{other wise} \end{cases}$$

$\widehat{1}$ is a vector whose entries are all ones

w_i is the cost associated with the installation of a PMU at particular bus i.

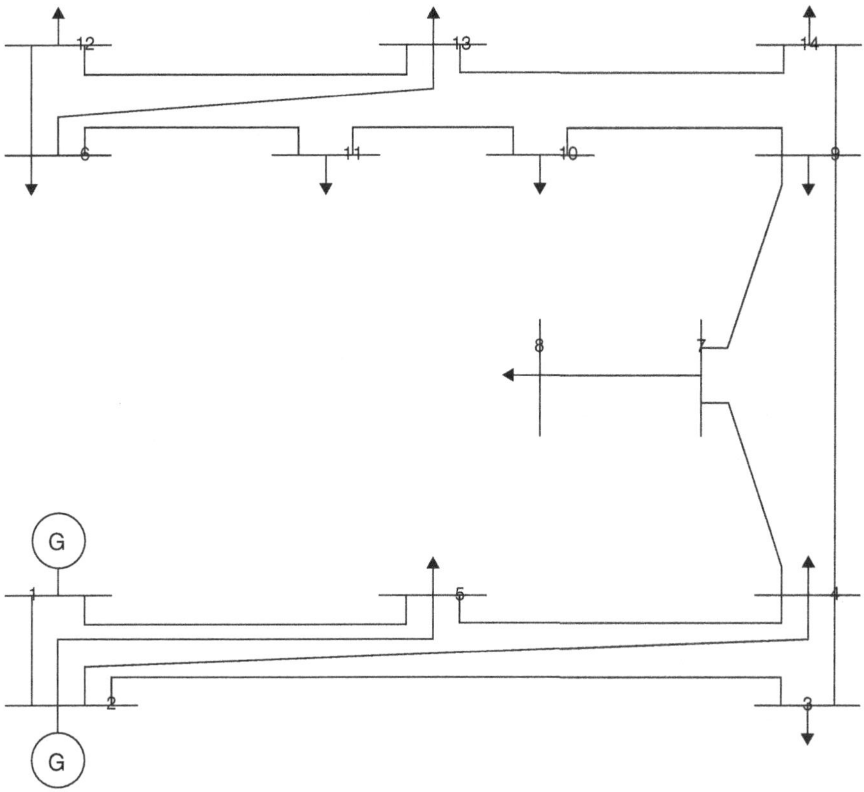

Fig. 10.1 IEEE 14-bus system. (Bus-7 is a zero-injection bus)

Figure 10.1 shows a single-line diagram of an IEEE 14-bus system. There are five generation points at buses 1, 2, 3, 6, and 8. Of them, only the generators at buses 1 and 2 create both active and reactive power while the generators at buses 3, 6, and 8 create reactive power. There are 11 load points connected at buses 2, 3, 4, 5, 6, 9, 10, 11, 12, 13, and 14 that consume the active and reactive power. With an ILP approach, we consider two possible cases to evaluate the IEEE 14-bus system: standalone PMU measurements (A), and a PMU with zero-injection buses (B). We use the System Observability Redundancy Index (SORI) proposed in [DDKS08] to choose the best optimal solution. The following example illustrates the ILP approach for the PMU placement problem with the two cases.

10.2.1 System With No Conventional Measurements and Zero/Flow Injections

A binary connectivity matrix, A, is formed to indicate links between the buses. The entries for matrix A are defined as follows:

$$A_{k,m}\begin{cases} 1, \text{if } \; k = m \\ 1 \; \text{ if } k \text{ and } m \text{ are connected} \\ 0 \; \text{if} \qquad\qquad \text{otherwise.} \end{cases}$$

Matrix $A_{k,m}$ can be directly obtained from the bus-admittance matrix by transforming its entries into binary form as follows:

$$A_{k,m} = \begin{bmatrix} 1 & 1 & 0 & 0 & 1 & 0 & 0 & 0 & 0 & 0 & 0 & 0 & 0 & 0 \\ 1 & 1 & 1 & 1 & 1 & 0 & 0 & 0 & 0 & 0 & 0 & 0 & 0 & 0 \\ 0 & 1 & 1 & 1 & 0 & 0 & 0 & 0 & 0 & 0 & 0 & 0 & 0 & 0 \\ 0 & 1 & 1 & 1 & 1 & 0 & 1 & 0 & 1 & 0 & 0 & 0 & 0 & 0 \\ 1 & 1 & 0 & 0 & 1 & 0 & 0 & 0 & 0 & 0 & 0 & 0 & 0 & 0 \\ 0 & 0 & 0 & 1 & 0 & 1 & 0 & 0 & 0 & 0 & 1 & 1 & 1 & 0 \\ 0 & 0 & 0 & 0 & 0 & 0 & 1 & 1 & 1 & 0 & 0 & 0 & 0 & 0 \\ 0 & 0 & 0 & 1 & 0 & 0 & 1 & 1 & 0 & 0 & 0 & 0 & 0 & 0 \\ 0 & 0 & 0 & 0 & 0 & 0 & 1 & 0 & 1 & 1 & 0 & 0 & 0 & 1 \\ 0 & 0 & 0 & 1 & 0 & 0 & 0 & 0 & 1 & 1 & 1 & 0 & 0 & 0 \\ 0 & 0 & 0 & 0 & 0 & 1 & 0 & 0 & 0 & 1 & 1 & 0 & 0 & 0 \\ 0 & 0 & 0 & 0 & 0 & 1 & 0 & 0 & 0 & 0 & 0 & 1 & 1 & 0 \\ 0 & 0 & 0 & 0 & 0 & 1 & 0 & 0 & 0 & 0 & 0 & 1 & 1 & 1 \\ 0 & 0 & 0 & 0 & 0 & 0 & 0 & 0 & 1 & 0 & 0 & 0 & 1 & 1 \end{bmatrix}$$

Here, a zero (0) in the connectivity matrix that is derived from the admittance matrix implies that there is no connection between busses k and m while a one (1) implies that there is a connection. The ILP constraints for this matrix can be formed as follows:

$$\text{Bus } 1 : f_1 = x_1 + x_2 + x_5 \geq 1 \tag{10.1}$$

$$\text{Bus } 2 : f_2 = x_1 + x_2 + x_3 + x_4 + x_5 \geq 1 \tag{10.2}$$

$$\text{Bus } 3 : f_3 = x_2 + x_3 + x_4 \geq 1 \tag{10.3}$$

$$\text{Bus } 4 : f_4 = x_2 + x_3 + x_4 + x_5 + x_7 + x_9 \geq 1 \tag{10.4}$$

$$\text{Bus } 5 : f_5 = x_1 + x_2 + x_4 + x_5 \geq 1 \tag{10.5}$$

$$\text{Bus } 6 : f_6 = x_6 + x_{11} + x_{12} + x_{13} \geq 1 \tag{10.6}$$

$$\text{Bus 7}: f_7 = x_4 + x_7 + x_8 + x_9 \geq 1 \tag{10.7}$$

$$\text{Bus 8}: f_8 = x_7 + x_8 \geq 1 \tag{10.8}$$

$$\text{Bus 9}: f_9 = x_4 + x_7 + x_9 + x_{10} + x_{14} \geq 1 \tag{10.9}$$

$$\text{Bus 10}: f_{10} = x_9 + x_{10} + x_{11} \geq 1 \tag{10.10}$$

$$\text{Bus 11}: f_{11} = x_6 + x_{10} + x_{11} \geq 1 \tag{10.11}$$

$$\text{Bus 12}: f_{12} = x_6 + x_{12} + x_{13} \geq 1 \tag{10.12}$$

$$\text{Bus 13}: f_{13} = x_6 + x_{12} + x_{13} + x_{14} \geq 1 \tag{10.13}$$

$$\text{Bus 14}: f_{14} = x_9 + x_{13} + x_{14} \geq 1 \tag{10.14}$$

The "+" operator in the formulation is a logical "OR" operation. The inequality operator ($\geq$) ensures that at least one variable in the sum will be non-zero. For example, consider the following constraint associated with buses 1 and 2:

$$\text{Bus 1}: f_1 = x_1 + x_2 + x_5 \geq 1$$
$$\text{Bus 2}: f_2 = x_1 + x_2 + x_3 + x_4 + x_5 \geq 1$$

Constraint $f_1 \geq 1$ implies that at least one PMU must be placed on either of the busses: $1(x_1)$, $2(x_2)$, and $5(x_5)$. Similarly, $f_2 \geq 1$ implies that at least one PMU should be installed for any one of the buses (1, 2, 3, 4, or 5) in order to make bus 2 observable or f_2 satisfied.

10.2.2 System With Zero-Injection Measurements

Zero-injection buses are buses where no currents are injected into the system. In other words, the buses have neither a load nor a generation unit. The use of Kirchhoff's Current law (KCL) to include zero-injection buses when modeling the OPP problem helps to further reduce the number of required PMUs. Each zero-injection node in a system creates an additional constraint in the formulation. Thus, the number of PMUs required for complete system observability can be reduced. For zero-injection bus i, let A_i indicate the set of buses that are adjacent to bus i. Let $B_i = A_i \cup \{i\}$. Let the number of zero injections be Z. Because there would not be any change in the current values with a zero-injection bus, the variables related to the zero-injection bus are eliminated from all constraints. The model formulation allows the constraints in an ILP framework to selectively allow some buses to be *unobservable*.

However, there are additional constrictions that need to be included for a fully observable system:

1. The unobservable buses must be adjacent to the zero-injection buses.
2. The number of unobservable buses in set B_i (zero-injection bus and its adjacent buses) is, at most, 1 (one).

Then, the LP formulation is written as follows:

$$\min \sum_{i=1}^{n} x_i$$

$$\text{s.t. } Ax \geq$$

$$u_j = 1 \forall j \notin B_1 \cup B_2 \ldots \cup B_z$$

$$\sum_{k \in B_i} u_k \geq |A_i| \forall i \in Z$$

In the IEEE 14-bus system shown in Fig. 10.1, bus 7 is a zero-injection bus. Thus, $Z = \{7\}$, *set* $A_7 = \{4, 8, 9\}$, and *set* $B_7 = \{4, 7, 8, 9\}$. The new constraint can be written as follows:

$$u_4 + u_7 + u_8 + u_9 \geq 3 \tag{10.15}$$

Constraints (1–15) indicate that at least three of the four buses (4, 7, 8, and 9) are required to keep the system fully observable. Thus, by including the above constraints, we use zero-injection variables, thereby reducing the number of PMUs. To our knowledge, zero-injection constraints have not been applied to a large case test system such as the IEEE 300-bus system. Our LP formulation includes results for the OPP problem by considering the zero-injection constraints for the 300-bus system.

10.2.3 System Observability Redundancy Index (SORI)

If there are multiple optimal solutions in an OPP placement problem, it is a challenge to identify the correct optimal solution. To solve this problem, we use the SORI index [DDK08]. If bus i is observed by a PMU n_i times, then the SORI (γ) is given by equation

$$\gamma_i = \sum n_i.$$

Choosing an optimal solution will provide maximum redundancy, enabling complete observability. This reliable, large value for the redundancy index is useful in contingency scenarios rather than choosing an optimal solution that are less redundant. In general, a larger redundancy value yields higher index values for the SORI. We believe that this SORI index is a good measure for our OPP placement problem.

10.2.4 Optimal Redundancy Criterion (ORC)

Although the OPP problem yields solutions for full observability, there is a likelihood that uncertainty during contingency situations will result in partial or zero observability for the system. This means that, in certain cases, it is better to include a certain percentage of buses that will be observed by multiple PMUs. Thus, the loss of PMU measurements from one device would not affect the observability of any other bus. The factor of percentage or observability weights of buses can be considered in different ways. Some studies consider that high-voltage buses are observed more than other buses while other researchers consider all generating buses to be weighed or observed more, considering them critical buses. We propose a new index, the Optimal Redundancy Criterion (ORC), which is based on the number of connections to each bus. The OPC is defined as the ratio of the total number of connections between buses to the total number of buses in the system.

$$\mathrm{ORC} = \frac{\text{Sum of all bus connections to each bus}}{\text{Number of buses}}$$

We can select the buses that have more connections or links by adapting the following criteria:

$$\mathrm{ORC} = \frac{\sum_{i=1}^{n} \sum_{j=1, i \neq j}^{n} B_{i,j}}{n} \qquad (10.(16))$$

where i and j represent the buses, $B_{i,j}$ represents the connections between i and , and n is the total number of buses in the system.

Let us consider constraints (1) to (14), where each constraint represents a bus in the IEEE 14-bus system. Here, the total number of connections for all 14 buses is the sum of their connections. For example, there are 56 total link connections between all buses.

$$B_1 + B_2 + \ldots B_{14} = 56$$

There are 14 buses in the test system.

$$\mathrm{ORC} = \frac{56}{14} = 4$$

Therefore, we increase the weights on the buses which have more connections than the ORC value. The buses with increased weights are buses 2, 4, and 9. The constraint formulations' weights appear on the right-hand side of the equation.

For full observability, we place a 1 on the right side of the constraint for a critical bus, implying that each bus is observed at least once. Based on the system requirements about how critical buses are, one can adjust the observability weights. For our test system, we chose 2 for the constraint.

10.3 IEEE Bus Test Cases and Simulation Results

We consider several test systems, such as the IEEE 14-bus system, the IEEE 30-bus system, the IEEE 57-bus system, the IEEE 118-bus system, the IEEE 300-bus system, and the Southern Region Indian Power Grid [GCR13], to validate the ILP method. The OPP programming is done with AMPL [AMPL].

Table 10.1 provides information about the zero-injection buses for each IEEE bus system.

Linear-programming (LP) formulations are created for different test systems with and without zero-injection buses. Table 10.2 shows the results obtained for all bus systems. We also performed a comparative analysis with other existing methods that are found in the literature. The results for the IEEE 57- and 118-bus systems are better when zero-injection buses are considered. The SRIPG system has a reduction of three PMUs when compared with the existing literature. To our knowledge, no studies have considered zero injections w.r.t. the IEEE 300-bus system. Our proposed ILP method proves to be more effective in minimizing the number of PMUs by considering zero-injection buses.

Table 10.1 Zero-injection buses

Bus systems	Zero-injection bus numbers	Number of zero-injection buses
14	{7}	1
30	{6, 9, 11,25,28}	5
57	{4,7,11,21,22,24,26,34,36,37,39,40,45,46,48}	15
118	{5, 9, 30,37,38,63,64,68,71,81}	10
300	{4,7,12,16,19,24,34,35,36,39,42,45,46,60,62,64,69, 74,78,81,85,86,87,88,100,115,116,117,128,129,130, 131,132,133,134,144,150,151,158,160,164,165,166,168, 169,174,193,194,195,210,212,219,226,237,240,244, 1201,2040,9001,9005,9006,9007,9012,9023,9044}	65

Table 10.2 Optimum number of PMUs

No.	Test systems	Minimum no. of PMU's			
		Without zero injections	With zero injections	Without zero injections	With zero injections
		ILP method		Other methods	
1.	IEEE 14-bus system	4	3	4	3
2.	IEEE 30-bus system	10	7	10	7
3.	IEEE 57-bus system	17	10	17	11
4.	IEEE 118-bus system	32	27	32	28
5.	SRIPG 208-bus system	55	NA	58	NA
6.	IEEE 300-bus system				NA

Table 10.3 System observability redundant index (SORI)

S. No.	Bus system	SORI			
		Without zero injections	With zero injections	Without zero injections	With zero injections
		ILP method		Other methods [DD08]	
1.	IEEE 14-bus system	18	14	19	15
2.	IEEE 30-bus system	51	46	NA	NA
3.	IEEE 57-bus system	72	52	72	61
4.	IEEE 118-bus system	167	182	164	152
5.	SRIPG 208-bus system		NA	NA	NA
6.	IEEE 300-bus system			NA	NA

Table 10.3 shows the results with the SORI index (γ) for the bus systems. As stated earlier, we will be able to further reduce the number of PMUs by considering zero injections. Our SORI results indicate that there is an increased number of PMUs, implying that the observability redundancy of the buses has increased, when compared with other methods. This implies that the chances of keeping the buses' observability are much greater during the contingency scenarios when using our approach.

During contingency situations, it is important to keep the observability alive for the critical buses. Using our ORC index, utilities can identify the critical buses and follow our placement solution. We strongly believe that our approach will provide redundancy to the critical buses with equal or least increase in the PMUs. Columns 2 and 3 in Table 10.4 show the number of PMUs that are required for full observability with the normal process and using ORC, respectively. Column 4 provides the value of the PMU's percentage increase (if any) when ORCs are considered. The rule that we apply here for the ORC index is that the critically identified buses should be observed by two PMUs. This means that, if one PMU's measurement is lost due to a failure, the critical buses should be observed by other PMUs in the system. Column 3 provides the number of PMUs required when the critical buses (formulated based on the ORC) are observed twice. By comparison, we

Table 10.4 Comparison of full observability and redundant observability

Bus system	Number of PMUs for full observability	Number of PMUs with ORC	% Increase in PMUs with ORC
IEEE 14-bus system	4	4	0
IEEE 30-bus system	10	10	0
IEEE 57-bus system	17	19	11.764%
IEEE 118-bus system	32	37	15.625%
SRIPG 208-bus system	55	60	7.272%
IEEE 300-bus system	[illegible]	[illegible]	[illegible]

Table 10.5 Estimated costs for the PMU equipment

Type of equipment	Cost (US$)
PMU with protection, automation, and controls	~15,000
PDC (up to 40 PMUs)	~ 8000
Synchronization clocks	~ 2000
Digital equipment (firewalls...)	~ 5000

Table 10.6 Computation time to obtain optimal solutions using the ILP method

Test bus systems	Proposed ILP method (ms)	Other methods [SSP12] (ms)
14 bus	0.00	660
30 bus	15.6	830
57 bus	15.6	870
118 bus	15.6	1340
208 bus	15.6	–
300 bus	15.6	–

conclude that there is a minimal or no increase in the number of PMUs when the ORC rule is enforced. The percentage increase w.r.t. different bus systems is seen in column 4.

In the real world, a utility needs to set up a number of things before it can place PMUs on the grid. These include constructing the communication infrastructure, procuring PMUs, Phasor Data Concentrators (PDC), Synchronization clocks, etc. Our related cost estimate to install a synchrophasor are as follows (Table 10.5):

The total cost run to ~ US$30,000 excluded from infrastructure, operational and labor costs. However, PDCs collect data from a number of PMUs. Therefore, a small utility can only install a few or a single PDC if it has fewer than 40 PMUs considering the PDC cost as a fraction of the number of PMUs.

We also measure our LP run time by measuring the computational run time taken to run the formulations. Table 10.6 shows the time taken by our ILP method using AMPL. The results are compared with the findings in the existing literature. We observe that it takes virtually zero seconds to compute the optimal solutions for the test cases while the other methods take some time to find the optimal solution. All the test cases are run on a PC with the following configurations: Intel Xeon W3503 processor @2.4 GHz and 8 GB of installed memory.

10.4 Conclusion

A PMU placement problem that considers zero injections using Integer Linear-Programming formulations is proposed. An ORC index is proposed to make the system fully observable with large redundancy. Additionally, placing synchrophasors creates economic benefits for smaller utilities using our approach. Our results indicate that the proposed approach is 100% reliable and scalable.

Chapter 11
Unbiased Optimal Power Flow (OPF) for Power Systems with Wind-Power Generation

Wind-power plants that are connected to power systems are often unable to utilize all available power due to transmission constraints. This chapter illustrates the situation with an optimal power flow (OPF) problem to schedule power from two different wind plants in a 6-bus system. A cost function is proposed to schedule wind power in an unbiased manner.

11.1 Introduction

Optimal power flow (OPF) is an enhanced form of the economic dispatch (ED) optimization problem where both active and reactive powers are treated as control variables and are adjusted to minimize the generation cost while observing the associated power-flow constraints. The constraints include, but are not limited to, thresholds on bus voltages, reactive power generation, and thermal ratings of transmission lines. The large-scale penetration of wind generation has created major technical challenges for power-system operation which, in turn, manifests itself in the associated ED and OPF problems for the concerned power system [XCF09]. This chapter discusses an OPF problem for a power system consisting of thermal and wind-turbine generators (WTG), and proposes a cost function to promote the equitable scheduling of power from multiple wind-power plants.

11.2 Motivation

One of the key challenges faced when integrating wind farms with the grid is spillage due to transmission constraints. A case study [NCK09] has shown that as much as 30% of the available wind may spill during periods of high wind-power availability without using reactive power-compensation technologies such as

P. Ranganathan, K.E. Nygard, *Distributed Linear Programming Models in a Smart Grid*, Power Electronics and Power Systems, DOI 10.1007/978-3-319-52617-1_11

flexible AC transmission systems (FACTS). While spillage can be reduced by upgrading the grid infrastructure or by introducing energy storage, economic and societal factors often impede such solutions. The power system's admittance properties may also cause more spillage in certain plants compared to others. Hence, a more nuanced approach is required to provide unbiased treatment for all power-plant operators.

11.3 OPF with Wind Generation

The standard OPF problem has been discussed in detail [FR12]. The optimization problem can be stated as determining the Day-Ahead dispatch for a system consisting of m buses, n_t thermal plants, and n_w wind plants. The objective is to minimize the cost given by (11.1), the cost summation due to scheduled wind-power generation ($P_{wind,j,i}$), thermal-power generation ($P_{therm,j,i}$), and transmission loss ($P_{loss,i}$) over N time blocks. The length of each block is 24/N hours.The thermal-generation cost is given by a polynomial cost function (C_t), while R_{loss} is the cost per unit of energy lost in the transmission. Equation (11.2) is the overall active-power balance constraint while the AC power flow is shown in Eqs. (11.3) and (11.4). Here, admittances between the kth and lth buses ($G_{k,l}+jB_{k,l}$) are taken in rectangular coordinates while voltage at the kth bus ($V_{k,i}\lfloor\delta_{k,i}$) is in polar coordinates. Physically, they convey

$$\min \sum_{i=1}^{N}\sum_{j=1}^{n_t} C_t\left(P_{\text{therm},j,i}\right) + \sum_{i=1}^{N}\sum_{j=1}^{n_w} C_w\left(P_{\text{wind},j,i}\right) + R_{\text{wholesale}}\sum_{i=1}^{N} P_{\text{loss},i} \tag{11.1}$$

$$\sum_{j=1}^{n_t} P_{\text{therm},j,i} + \sum_{j=1}^{n_t} P_{\text{wind},j,i} + P_{\text{loss},i} + \sum_{j=1}^{m} P_{\text{load},j,i} = 0 \tag{11.2}$$

$$P_{\text{gen},k,i} - P_{\text{load},k,i} = V_{k,i}\sum_{l=1}^{m} V_{l,i}(G_{k,l}\cos(\delta_{k,i}-\delta_{l,i}) + B_{k,l}\cos(\delta_{k,i}-\delta_{l,i})) \tag{11.3}$$

$$Q_{\text{gen},k,i} - Q_{\text{load},k,i} = V_{k,i}\sum_{l=1}^{m} V_{l,i}(G_{k,l}\sin(\delta_{k,i}-\delta_{l,i}) + B_{k,l}\sin(\delta_{k,i}-\delta_{l,i})) \tag{11.4}$$

that the active ($P_{gen,k,i}$) and reactive ($Q_{gen,k,i}$) powers generated in the kth bus minus the load at that bus ($P_{load,k,i}, Q_{load,k,i}$) is equal to the net power flowing from the bus. Active and reactive powers for thermal generation are constrained by the generator ratings shown in Fig. 2.1. The reactive power limit for wind generation is calculated based on a power factor of 0.95 lag/lead [QH08]. Voltage limits for all buses are set between 0.9 and 1.1 p.u. The generation ramp rate limits and transmission-line capacity constraints have not been considered.

11.4 Cost Function for Unscheduled Wind

Because there is no fuel cost associated with wind energy, power scheduled from a wind plant cannot be assigned a cost in the traditional sense. At high levels of wind-capacity penetration, the system's operational costs can increase because of the wind's variable nature [SMD09]. The levelized cost of energy (LCOE), which is calculated using a net present value analysis, is an indicator of the wind energy's generation cost [Sat06] and is calculated using Eq. (11.5). The LCOE is used as a performance index in the coordinated output control of multiple DG resources in [JTM10].

$$C_{\text{wind}}^{\text{LCOE}} = \frac{C_{\text{wind}}^{I}}{8760n}\left(\frac{1}{P_{\text{wind}}^{\text{rated}} CF_{\text{wind}}^{\text{design}}}\right)\left\{1 + M\left[\frac{(1+D)^n - 1}{D(1+D)^n}\right]\right\} \tag{11.5}$$

From Eq. (11.5), it can be seen that, for a plant of rated capacity $\left(P_{\text{wind}}^{\text{rated}}\right)$ with fixed investment parameters such as the amount of capital invested $\left(C_{\text{wind}}^{I}\right)$, the discount rate ($D$), the plant's lifetime (n), and the annual maintenance-cost fraction (M), the wind plant's design capacity factor $\left(CF_{\text{wind}}^{\text{design}}\right)$ heavily influences the LCOE. In order to enable a wind plant to be financially viable, enough power must be scheduled so that the plant can maintain an average capacity factor $\left(CF_{\text{wind}}^{\text{actual}}\right)$ that is close to the design capacity factor. This chapter calculates the opportunity cost for unscheduled wind power (C_{w}) using Eq. (11.6) based on the difference between the actual and design LCOE $\left(\Delta C_{\text{wind},j}^{\text{LCOE}}\right)$ as given by Eq. (11.7). The authors assume that the short-term estimated wind power $\left(P_{\text{wind},i}^{\text{pred}}\right)$ is available for the OPF problem and comes from physical and/or statistical models. A good overview of short-term wind-power prediction methodologies is presented in [Gie11]. Additionally, in order to quantify the amount of predicted wind that is utilized by wind plants, a wind-utilization factor (U_{wind}), given by Eq. (11.8), is defined.

$$C_{\text{w}}\left(P_{\text{wind},j,i}\right) = \Delta C_{\text{wind},j}^{\text{LCOE}}\left(P_{\text{wind},i}^{\text{pred}} - P_{\text{wind},j,i}\right) \tag{11.6}$$

$$\Delta C_{\text{wind},j}^{\text{LCOE}} = \left(\frac{1}{CF_{\text{wind},j}^{\text{actual}}} - \frac{1}{CF_{\text{wind}}^{\text{design}}}\right)\frac{C_{\text{wind},j}^{I}}{P_{\text{wind},j}^{\text{rated}} 8760n}\left\{1 + M\left[\frac{(1+D)^n - 1}{D(1+D)^n}\right]\right\} \tag{11.7}$$

$$U_{\text{wind}} = \frac{\sum_{i=1}^{N} P_{\text{wind},i}}{\sum_{i=1}^{N} P_{\text{wind},i}^{\text{pred}}} \tag{11.8}$$

11.5 Case Studies and Results

The 6-bus system shown in Fig. 11.1 is used in this study. The bus-admittance matrix is constructed using data from [Web97] where each line has an impedance of 0.04 + 0.08j ohm and a line-charging reactance of 0.02j ohm. The load profile shown in Fig. 2.2 is used, and the reactive load is calculated by assuming a lagging power factor of 0.85 for all loads.

In the first scenario, we solve the OPF problem to minimize the thermal plant's generation cost and transmission loss; i.e., the wind-power cost function is not included. However, power generation is scheduled for N time blocks for all the

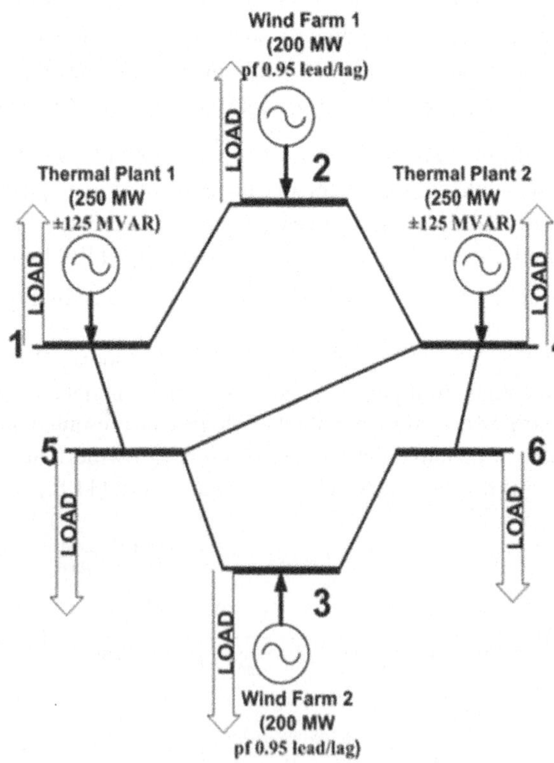

Fig. 11.1 Six-bus power system (Ward-Hale model) [Web97]

plants. The problem is solved for three different, predicted wind-power profiles, and the results are tabulated in Table 11.1. The OPF solution consistently allows the wind plant connected to bus 3 to utilize more of its predicted power compared to the plant connected to bus 2. Another correlation is that a higher wind-power availability (predicted) does not translate to higher wind utilization.

In the second scenario, we include the cost function defined by Eqs. (11.6) and (11.7). We assume that the $CF_{\text{wind}}^{\text{actual}}$ for the wind plants at buses 2 and 3 are 25 and 35%, respectively. This amounts to a $\Delta C_{\text{wind}}^{\text{LCOE}}$ of \$10.3/MWh and \$2.45/MWh, respectively, for the two wind plants. The load and wind profiles remain the same as in the previous scenario. The results are tabulated in Table 11.2.

The OPF solution schedules more power for the wind plant at bus 2 in all cases. Interestingly, the contribution of wind energy toward meeting the total energy demand remains nearly the same irrespective of the cost functions. However, the losses are comparatively higher with the wind-cost function. The thermal plants are scheduled in such a way that the most efficient plant has the highest capacity factor.

The wind power scheduled for each time block in scenario 2 using wind profile 1 is shown in Fig. 11.2. The voltage profile at the load buses generated by the OPF

Table 11.1 OPF solution for the first scenario

Wind-power profile	1	3	2
Average predicted wind power (MW)	159.2	136.7	113.3
Wind utilization WTG1 (%)	70.7	72.4	86.1
Wind utilization WTG2 (%)	97.6	1	1
Total wind contribution (%)	61.3	53.9	48.2
Energy loss (MWh)	294	287.5	303.7
Load bus 1's minimum voltage (p.u.)	1.01	1.02	1.01
Load bus 2's minimum voltage (p.u.)	0.97	0.97	0.97
WTG1's minimum voltage (p.u.)	1.1	1.1	1.11
WTG2's minimum voltage (p.u.)	1.01	1.03	1.02

Table 11.2 OPF solution with the wind-power cost function

Wind-power profile	1	2	3
Average predicted wind power (MW)	159.2	136.7	113.3
Wind utilization WTG1 (%)	1	93.7	100
Wind utilization WTG2 (%)	70.3	79.4	87
Total wind contribution (%)	61.5	53.9	48.2
Energy loss (MWh)	369	317	327
Load bus 1's minimum voltage (p.u.)	1.01	0.98	1.01
Load bus 2's minimum voltage (p.u.)	0.97	0.92	0.97
WTG1's minimum voltage (p.u.)	1.11	1.05	1.11
WTG2's minimum voltage (p.u.)	1	1.02	1.01
Thermal plant 1 CF (%)	30	30	30
Thermal plant 2 CF (%)	37.8	50.6	60.6

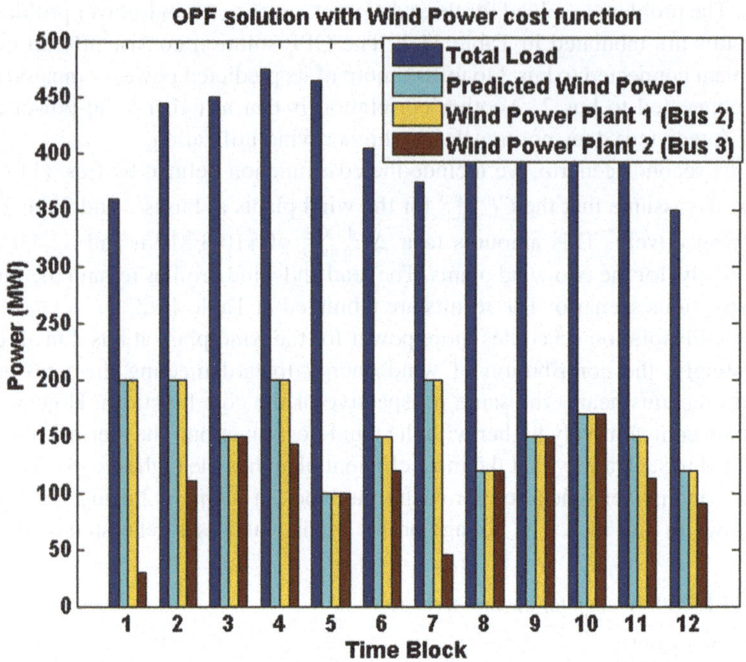

Fig. 11.2 Comparing power scheduled from both wind-power plants

solution with and without the wind-power cost is shown in Fig. 11.3. The load voltages are slightly less in certain time blocks for the OPF solution when the wind-power cost function is used compared to the result without that cost function.

11.6 Further Discussion

In order to illustrate the merit of the proposed cost function for wind power, the cost function proposed in [HYB08] is reproduced in its original form as shown in Eq. (11.9). This function penalizes unscheduled wind-power production and is part of a comprehensive cost function for an economic dispatch problem.

$$C_{p,w,i}(W_{i,av} - w_i) = k_{p,i}(W_{i,av} - w_i) \tag{11.9}$$

In Eq. (11.9), the term $W_{i,av} - w_i$ is the difference between the available and scheduled wind power while $k_{p,i}$ is the penalty-cost coefficient.

It can be observed that Eq. (11.9) is similar to Eq. (11.6) except for a key difference. The term $k_{p,i}$, which was an arbitrary value in [HYB08], has been replaced by $\Delta C_{\text{wind}}^{\text{LCOE}}$ which is calculated based on the actual capacity factors of the wind plants.

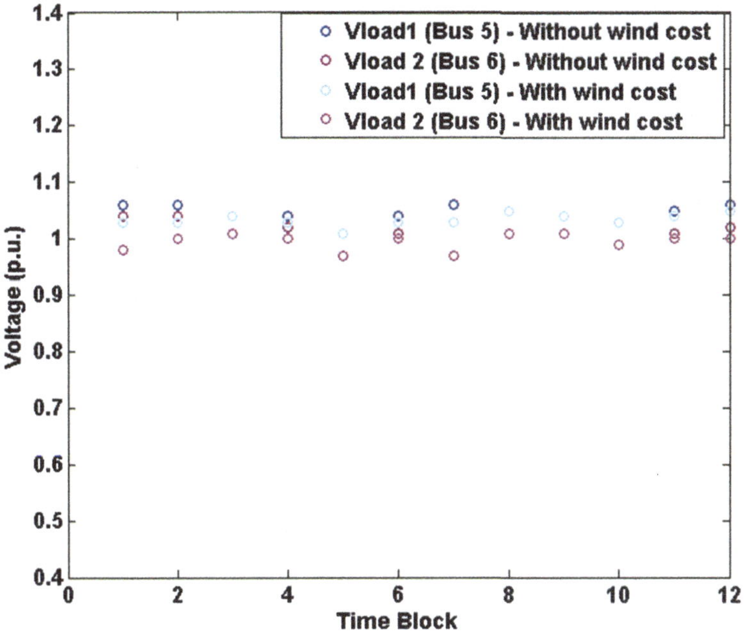

Fig. 11.3 Voltage profile of the load buses (buses 5 and 6)

11.7 Conclusion

An OPF optimization problem for power systems with thermal- and wind-power plants was investigated. A formulation to calculate the opportunity cost for unscheduled, predicted wind-power production was proposed and compared with the existing literature. Results from case studies with a six-bus system were presented. The work may be extended to systems with more buses and additional constraints.

Chapter 12
Smart-Grid Optimization Using A Capacitated Transshipment Problem Solver

12.1 Introduction

Creating an autonomous, self-healing electrical grid is one of the most important challenges facing electric-energy providers. Such a system, known as the "smart grid," must interweave a multitude of systems, both software and hardware, in order to form a complete solution that is capable of meeting the requirements outlined by the United States Department of Energy (DOE). According to the DOE [LSC05], "It is a colossal task, but it is a task that must be done."

This work focuses on a single aspect of those systems: optimal electrical flow through the smart-grid network as determined by a cost factor. The cost factor can be a different performance metric for various optimization objectives, including the distance between generators and customers, electric-line repair times, or failure rates.

In order to determine the best solution for multiple, different cost-related problems that are associated with the smart grid, the capacitated transshipment problem, or CTP, was chosen from the mathematical field of linear programming to model the smart-grid network and its values. Using this model, a custom CTP Solver was developed, allowing users to easily determine the optimal network flow for a given smart-grid network topology.

For a better understanding about the purpose of the CTP Solver and this work, some background information on the smart grid and linear programming is helpful. In particular, the smart grid, the simplex algorithm, the network-flow problems, and, especially, the capacitated transshipment problem are reviewed in some detail.

P. Ranganathan, K.E. Nygard, *Distributed Linear Programming Models in a Smart Grid*, Power Electronics and Power Systems, DOI 10.1007/978-3-319-52617-1_12

12.2 Smart Grid

"The grid," refers to the United States' electric power grid. It is a network consisting of substations, transformers, and transmission lines that are used to provide electricity to homes and businesses; that electricity comes from an electric power plant [DOE]. The existing electric grid originated in the 1890s; it has grown and evolved to a network of more than 9200 electric-generating units. These generators are capable of producing over one million megawatts of generating capacity, which is then made available through more than 300,000 miles of transmission lines [DOE].

While the current grid is considered an engineering marvel [DOE], it is beginning to show its age. As electricity needs and demands increase and advance so, too, must the electric grid that provides the power. It follows, then, that "smart grid" refers to using computer-based remote control and automation in an effort to modernize the utilities' electricity delivery systems [DES]. Among the many benefits for these automated systems is improving the electrical grid's reliability by dynamically re-routing power, as needed, to avoid cascading failures.

12.2.1 *Self-Healing System*

One of the greatest benefits of a fully functional smart grid is the concept of self-healing. Current methods of outage detection vary and can be primitive at best, requiring customers to call the electric provider with service-interruption notifications. This type of recovery solution is completely reactive and, often, much too slow to prevent catastrophic failures such as cascading outages. When a generator fails, a large system is affected and can cause other generators to overload. As stations continue to fail, the outage spreads farther throughout the network.

Self-healing in the smart grid is just one aspect of a larger concept that is referred to as "distribution intelligence." It is concerned with the utility's distribution system: the wires, switches, and transformers that connect the utility substation to the customers [DOE].

Outage detection is another aspect of smart-grid distribution intelligence. The CTP Solver assumes that an outage has been detected and concerns itself with the optimal redistribution of power based on the smart-grid network's current state (Fig. 12.1).

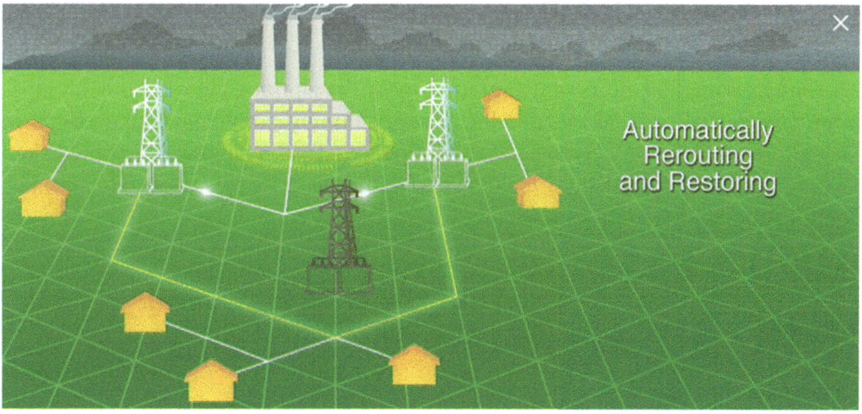

Fig. 12.1 Dynamic electrical power rerouting [DOE]

12.3 Linear Programming

Sometimes referred to as "linear optimization," linear programming can be defined as the general approach to modeling and solving linear mathematical models [CI94], p. 2.

The term "programming" can be a bit misleading because it does not specifically mean computer programming, which many people might assume at first glance. In this context, the term provides a more general reference to problem solutions; of course, these solutions could, in fact, be implemented as computer programs, but that is not a requirement.

Three basic steps are usually followed when formulating a model to represent a given linear programming problem:

1. Determining the decision variables
2. Formulating the objective function
3. Formulating the constraints

The decision variables represent measurable aspects of the problem, such as the unit cost. The objective function seeks to optimize the problem, and the constraints are limitation requirements. With a model in place, the key concept of linear programming is to optimize the objective function, which can also be thought of in terms of minimization or maximization. When feasible, the solution to a linear-programming problem is the best possible result of the objective function's value with respect to any constraints.

Some canonical examples of linear-programming problems include the assignment problem; the traveling salesman problem; and the transportation problem, a simplified variation of the capacitated transshipment problem. There are currently a number of software solutions that focus on solving linear-programming and optimization problems, including AMPL and SAS.

12.3.1 Simplex Method

The simplex method, sometimes referred to as the simplex algorithm, is an algebraic process that is used to solve linear-programming problems. George Bernard Dantzig is considered the creator of the simplex method, which was first published in 1947 and was detailed at a 1951 Cowles Commission for Research in Economics conference [Dan51].

To summarize its concepts, the simplex method mathematically models a problem so that its solution space can be described in one of three ways:

1. Feasible Solution
2. Infeasible Solution
3. Optimal Solution

An infeasible solution is any point that does not satisfy every constraint and non-negativity condition of the linear program.

A feasible solution is any point that satisfies every constraint and non-negativity condition of the linear program. The set of all feasible solutions is known as the feasible region; this is the equivalent of the intersection of all feasible solutions.

If the linear program has a bounded feasible region, meaning that the feasible solution space is fully contained, an optimal solution is some point on the feasible region's boundary. The simplex method effectively traverses the boundary in search of these optimal points, which are also referred to as extreme points.

12.3.2 Network-Flow Problems

In general, a network-flow problem is any from a particular class where the solution space can be described using nodes and arcs connecting those nodes with unit flow along the arcs transferred from one node to another.

One of the keys to developing an efficient algorithm for this class of linear programming problems is establishing a relationship between the algebraic and graphical representations of basic solutions. In particular, one of the most important relationships is the one that exists between basis matrices and rooted spanning trees [CI94], p. 320.

Theorem 9.1: Every rooted spanning tree is a basis [CI94], p. 320.

Theorem 9.3: Every basis is a rooted spanning tree [CI94], p. 322.

A basis is defined as a collection of vectors, $\mathbf{a_1}, \mathbf{a_2}, \ldots, \mathbf{a_k}$, in an n-dimensional (real) Euclidean space, denoted by R^n, where the following conditions hold:

1. $\mathbf{a_1}, \mathbf{a_2}, \ldots, \mathbf{a_k}$ span R^n.
2. If any of these vectors is deleted, the remaining collections of vectors do not span R^n [BJS10], pp. 48–49.

Recall that the standard form of a linear-programming problem can be written in the same vector notation and that an initial basis tree can be calculated using the Big-M method, allowing the simplex algorithm to be applied to network-flow problems.

Once the basis tree of a network-flow problem has been established, the network's simplex algorithm can be implemented. Network-flow problems are typically considered minimization problems, although they can be easily changed into maximization problems by using negative cost values and changing the sign polarity to positive values once the solution is obtained.

The four basic steps of the network's simplex algorithm are as follows:

1. Determine the primal and dual solutions.
2. Check for optimality.
3. Determine the departing variable.
4. Pivot and update.

Modified versions of these steps are detailed in Section 5.

12.4 Capacitated Transshipment Problem

The capacitated transshipment problem, or CTP, is an important network-optimization problem [BJS10], p. 513 that consists of four primary elements: supply nodes, demand nodes, transshipment nodes, and connective arcs. The basic concept of the CTP is to find a minimum cost path that connects every node of the network and transfers all units of flow from the supply nodes to the demand nodes without violating any of the network's arc capacities.

12.4.1 Transportation Problem

The capacitated transshipment problem can be easily understood through a simplified network-flow variation that is known as the transportation problem [CI94], p. 350. With the transportation problem, unit flow is pushed along the network arcs from the supply nodes to the demand nodes. All supply units in the network must be transferred from the supply nodes to the demand nodes. This is known as the flow-balance constraint and can be written as Eq. (12.1).

Due to this constraint, if the total supply of a transportation problem is not equal to its total demand, the problem is infeasible. The goal of the transportation problem is to find the basis tree that minimizes the total cost of the unit flow along the network.

$$\text{Flow Out - Flow In - Supply} = 0$$

Eq. 12.1 Flow-balance constraint

12.4.2 Transshipment Nodes

The capacitated transshipment problem generalizes the transportation problem by adding transshipment nodes; or nodes with zero supply or demand. This means that the nodes must still be connected in the basis tree, but all flow units simply pass through the node. Transshipment nodes can cause degenerate arcs (arcs with a unit flow of zero). These degenerate arcs have the potential to create infinite loops and must be handled properly to prevent cycling through the same solutions.

12.4.3 Arc Capacities

In addition to transshipment nodes, the capacitated transshipment problem also adds capacities and lower-bound requirements to arcs. Flow along any given arc must be at least as much as the lower bound and not more than its capacity. Violating either of these constraints causes an infeasible solution. Arcs with a flow that is equal to their capacity can be considered as part of the solution without taking up space in the basis tree. These arcs are referred to as non-basic arcs with bounded flow.

12.4.4 CTP Standard Form

The capacitated transshipment problem can be described in algebraic standard form as shown in Fig. 12.2. The objective function is to minimize the sum total of all arc unit flows multiplied by their costs.

Constraint (1) ensures the flow balance at every node by making sure that the total flow from a node is the same as the total flow into it with respect to the node's supply and demand requirements. This constraint also ensures that supply units are distributed from all supply nodes to all demand nodes, creating a zero-net unit flow for the entire network.

Constraint (2) ensures that all arcs have a non-negative unit flow.

Constraint (3) ensures that no arc capacities or upper-bound limits are violated.

Constraint (4) ensures that no arc lower-bound requirements are violated.

Formulating the problem in standard form allows the network's simplex algorithm to be applied in order to calculate the optimal solution.

Minimize $z = \Sigma c_{ij}x_{ij}$

Subject to

$x_{ji} - x_{ij} + b_i = 0$ for all arcs i,j
$x_{ij} \geq 0$ for all arcs i,j
$x_{ij} \leq u_{ij}$ for all arcs i,j
$x_{ij} \geq l_{ij}$ for all arcs i,j

where

c = arc cost
x = arc flow
u = arc upper bound
l = arc lower bound
b = node supply (negative value for demand)

Fig. 12.2 CTP standard form

12.4.5 CTP and the Smart Grid

Representing the smart grid using the model for a capacitated transshipment problem allows multiple cost- and network-flow-related problems to be solved easily. One of these problems is the smart grid's self-healing aspect.

When a critical failure is detected in the system, the CTP can be used to find an optimal and inherently feasible redirected path so that energy could be redistributed throughout the smart grid. By finding the best-alternative distribution path, customer outages can be minimized and rectified almost as quickly as they occur, when possible. Figure 12.3 shows the diagram for the IEEE 14-bus system and its network representation for use with the CTP Solver.

12.5 Implementation

Before getting into the nuts and bolts of the CTP Solver application, it is important to understand the motivation behind its design and architectural decisions. Once these design goals have been examined, it will, hopefully, be clear why certain development choices were implemented.

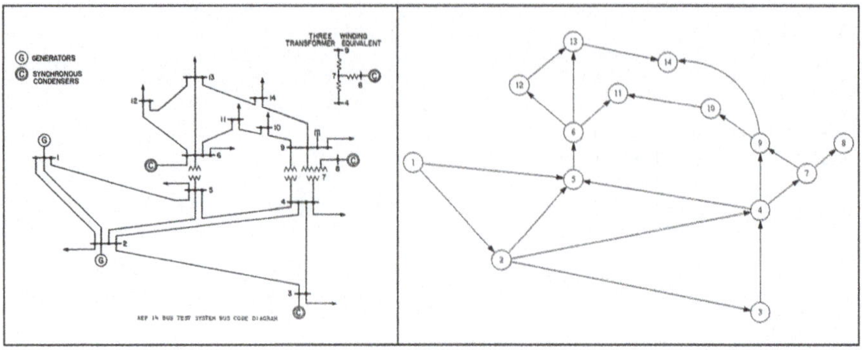

Fig. 12.3 *Left*: IEEE 14-Bus Test System diagram. *Right*: IEEE 14-Bus Test System network representation

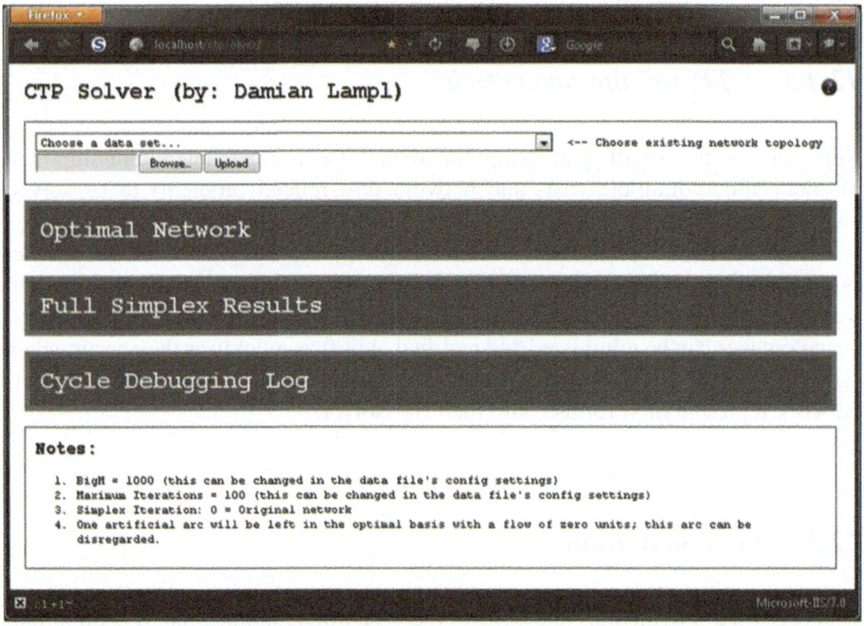

Fig. 12.4 CTP Solver (screenshot and light CSS)

12.5.1 Design Goals

The design goals of the CTP Solver can be broken into two primary categories:

1. User Experience
2. Application Development and Maintenance (Fig. 12.4)

12.5.1.1 User Experience

One of the primary factors driving the development decisions was the desire to make the application convenient and easy for its end users. Considerations were made to understand how the application would, generally, be used and what information would be beneficial to have available for display as well as download. The user-experience design goals can be summarized with a handful of basic conveniences that were implemented with the user in mind.

- Automated Initial Basis Calculation
  - User not burdened with calculating an initial basis to feed into the network topology.
- Bi-Directional Arc Capability
  - Flag allowing the flow to travel in either direction along a single arc.
- Multiple Network Topology Data Formats
  - HTML Table (display)
  - Database (import)
  - XML (import and export)
  - CSV (export)
- Decimal Values
  - Enter the information as it exists instead of requiring pre-calculation transformations into integer data types, etc.
- Configuration Options
  - Big M Value
  - Simplex Iterations Limit
  - Show Detailed Simplex Information
  - Show the Cycle Debugging Log
- Help System
- Accessibility
  - Web Application

12.5.1.2 Application Development and Maintenance

In addition to the user-experience design goals, careful considerations were made regarding the CTP Solver's development and ongoing maintenance. The purpose for the application-development and maintenance goals was to design the application's architecture and code with the developer in mind. The application-development and maintenance goals can be summarized by some basic developer-focused concepts.

- Ease of Use
  - Simplicity
  - Self-Documenting Code
  - Partial Classes
- Universal Applications
  - Generic Network Concepts
- Calculation Precision
  - Decimal Data Type
- Maintainability
  - Object-Oriented Concepts

12.5.2 Implementation Overview

The CTP Solver has been implemented as a C# ASP.NET web application, primarily for accessibility among its users. Many applications are written as standalone software; are platform dependent; or, in the case of an environment such as Java, are dependent on some other service that must be installed on the end user's machine.

While dependent on the ASP.NET framework on the host server, as a web application, the CTP Solver can be made available to any machine or mobile device that is connected to the internet, regardless of the operating system. This inherently widens the potential user-base and also minimizes the required technical capabilities for the people who would most benefit from its usage. The ability for the CTP Solver to be as generic as possible is as important as the usage ubiquity that is provided by the web. While it is a vital requirement to be able to handle smart-grid network problems, it is equally imperative that the CTP Solver can process any capacitated transshipment problem network. This requirement is met with basic supply, demand, flow, and cost concepts, among others.

When a user uploads a custom file or selects a network option from the dropdown of available choices, the CTP Solver reads the XML file (or database) and commits the topology to memory. The network can be fully described through two primary IList data structures: nodes and arcs; each one is described in more detail below.

Once the network is in the memory, the CTP Solver iterates through its modified simplex algorithm and displays the resulting optimal solution to the user. In addition to some basic computational data, such as the execution time of the entire process, the optimal network is also available for download in both the .CSV and

XML format. This allows the user to easily import the optimal network into other applications.

12.5.3 Smart-Grid Possibilities

A benefit of web applications related to the smart grid is that an implementation could be created in such a way that a single CTP Solver application could simultaneously calculate optimal solutions for multiple client-network configurations. This capability could be of some benefit when planning systems with budget constraints as well as providing a single source of maintenance for IT staff. Because the CTP Solver can connect to a database and can read XML files, it could be easily integrated with other smart-grid systems, such as failure-notification solutions, providing automatic, optimal electric-flow rerouting based on the supplied network topology of the available nodes and arcs. Because arc capacities are taken into consideration, the cascading-failure dynamic could possibly be avoided by ensuring that the network flow is feasibly rerouted.

As the smart-grid system grows, more cost-performance measures are likely to be revealed. Because the CTP Solver was built with generic concepts in mind, it should be able to calculate the optimal results for any new network topology that can be modeled by the capacitated transshipment problem. These potential smart-grid applications, and more, are made possible by the CTP Solver's integration of various custom and framework-native data structures.

12.5.4 Data Structures

Traditional linear-programming techniques implement primitive data structures for network-topology descriptions, such as arrays, or the standard model formulation for use in specific modeling software such as AMPL. While these methods make sense for application speed and simplicity, the CTP Solver introduces object-oriented practices to take advantage of robust, modern programming-language capabilities, such as LINQ for the ASP.NET framework, while maintaining very comparable speed on the current hardware.

1. **DataSet:** The DataSet is a native structure to the ASP.NET framework. The benefit of a DataSet object is that it takes on some properties of a traditional relational database, including concepts such as rows and columns. The CTP Solver reads the user-supplied network into a DataSet object. In doing so, the same architecture can be used to read from either XML or a traditional database, such as MSSQL or MySQL, eliminating code redundancy for, essentially, the same process.

2. **IList:** An IList is another native ASP.NET data structure that represents a collection of objects, with the "I" referring to the term "Iterative," making it an iterative list. Objects stored in an IList can be "queried" much like relational databases by using syntax that is similar to standard SQL. Another benefit of utilizing an IList collection object is that it only uses as much memory as it needs. Objects can be added and removed from the collection without requiring explicit dimensions. This makes resizing the structures much more efficient than resizing arrays.

   The primary limit of concern with the IList structure is the maximum number of elements allowed in a single IList object, which is the same limit as an array. Theoretically, the maximum number of elements in an IList is 2,147,483,646; however, a network of that size would most likely benefit from some kind of partitioning, such as a modified Dantzig-Wolfe decomposition approach [NP]. Using a small number of these structures, the CTP Solver is able to intuitively represent the entire network topology and more.
3. **LINQ:** "LINQ" stands for Language-Integrated Query [Mic15]. First introduced in Visual Studio 2008, LINQ allows strongly typed object collections (such as the IList described above) to be queried, providing an easy system to extract relevant information from the data. LINQ queries are used generously in the custom Arc class to provide partial lists for structures such as the basis tree and all non-basic arcs.
4. **Node:** A node is a custom object class that is created to represent each node in the network. All network nodes are added to an IList for easy access during the calculations. The properties of each node are described in Table 12.1.

   The Node class has a single method that is shown in Table 12.2.

5. **Arc:** An arc is the second custom object class that is used to represent connections, which are sometimes referred to as edges or links, between nodes. As with nodes, all network arcs are added to an IList data structure, allowing a simple representation of the network connections. The arc objects' properties are described in Table 12.3.
   In addition to the properties just described, the arc class has a few important methods that are used for the CTP Solver's algorithms, as described in Table 12.4.
6. **Algorithm Methods:** The modified simplex algorithm being implemented (described below) is broken into a handful of methods as shown in Table 12.5.
7. **Miscellaneous Data Structures:** In addition to the primary data structures, a small number of helper and utility classes are created to handle various aspects of the application. Three utility classes are used to handle commonly used functions, user-configurable options, and database interactions. In addition, all custom-class structures are created as partial classes. Using a partial class allows its methods to be defined in separate files. It is common practice to define an entire class in a single file using the class name as the file name. The benefit of using a partial class comes from the ability to logically separate categorized

Table 12.1 Node properties

Property	Data type	Description
id	integer	Providing each node with a unique id allows easy reference to individual objects and makes database interactions nearly seamless
ConnectedArcs	IList<Arc>	List of all arcs where the node is either a Head or Tail. This list is used to calculate the net flow entering and leaving the node in order to enforce the flow-balance requirements
Demand	decimal	A node's demand represents the number of flow units required at that node
Depth	integer	The node's depth is its level in the basis tree starting from the root node
Name	string	The name of the node is mainly included for the benefit of human readability if the network topology is printed to a screen or if future functionality includes generating network diagrams, etc. The CTP Solver does not use this property for any purpose in its algorithms other than simply storing the information
NetFlow	decimal	It is defined as the total flow coming into a node, minus the total flow leaving a node, plus the node's supply (which will be subtracted when representing the demand because it is then a negative value)
Parent	integer	The node's parent is used in the basis tree structure. It represents the node immediately connected to and one depth level above the current node. A negative parent value represents a reflected arc in the basis-tree structure
Potential	decimal	A node's potential is the equivalent to a dual variable in linear programming. This value is used to calculate the reduced cost of non-basic arcs in order to find the best candidate arc to enter the basis tree
Successor	integer	The node's successor is also used in the basis tree structure. It represents the node following the current node in the preorder thread. Unlike parent nodes, successor nodes in the preorder thread are not necessarily directly connected by an arc
Supply	decimal	The node's supply represents the number of flow units that are available from that node. For simplicity in the CTP Solver algorithm, the supply property also represents a given node's demand by reversing its polarity to a negative sign. Transshipment nodes are given a supply value of zero

Table 12.2 Node methods

Method	Description
GetArcs	Returns an IOrderedEnumerable list of arcs where the node is either a Head or Tail. This method is used to set the value of the ConnectedArcs property

methods of a class for easier maintainability as well as allowing common generic methods to be automatically generated.

To speed the development of creating the node and arc database interaction methods, a custom object-relationship mapping application (commonly referred

Table 12.3 Arc properties

Property	Data type	Description
Id	integer	Providing each arc with a unique id allows easy reference to individual objects and makes database interactions nearly seamless
BasisOrder	integer	Originally used before the initial basis was automatically calculated, the basis order represents the arc's order in the basis tree
Capacity	decimal	The capacity is the upper-bound limit on units of flow that can move across the arc at a given time. The capacity adheres to the Big M maximum value limit that is set by the user in the config element
Cost	decimal	The cost represents the price of moving one unit of flow across the arc. It is important to note that an arc's total cost is calculated by multiplying the arc cost and flow. The cost adheres to the Big M maximum value limit that is set by the user in the config element
Cpx	decimal	Cpx is the capacity's calculated value minus the flow of a given arc
Flow	decimal	The flow is the number of units pushed across the arc from the tail node to the head node. Its value must fall between the arc's capacity and lower bound
Head	integer	The arc's head is the id of its destination node. It is the stopping point of a directed arc. Unit flow along the arc starts at the tail node and moves toward the head node
IsArtificial	Boolean	The IsArtificial flag determines whether the arc is artificial or real. Artificial arcs are used when creating the initial basis and connecting real nodes to the single artificial node. If an arc is artificial, it is not allowed to re-enter the basis tree once it has been removed
IsBasic	Boolean	The IsBasic flag determines whether the arc is in the basis tree. It could be considered somewhat redundant due to the BasisOrder property, but it is used in some logic checks and output displays
IsBidirectional	Boolean	The IsBidirectional flag determines whether the arc can be considered to have the flow move in either direction: from the tail node to the head or from the head node to the tail
LowerBound	decimal	The lower bound is the minimum number of flow units required on the arc
ReducedCost	decimal	The reduced cost is a value to determine which non-basic arc will enter the basis. The arc with the best reduced cost, meaning that the arc that will lower the basis' overall cost by the largest amount, is chosen to enter the basis
SameCycleDirection	Boolean	The same cycle direction flag determines whether the arc follows the same flow direction as the cycle's entering arc
Tail	integer	An arc's tail is the id of its source node. It is the starting point of a directed arc

Table 12.4 Arc methods

Method	Description
GetArc	Uses a LINQ query to return an arc from the provided head and tail node id's
GetBasis	Uses a LINQ query to find all basic arcs in the provided IList object and represents the basis-tree structure. It returns an IOrderedEnumerable list of arcs that is used for procedures that only require the basis tree
GetChildArcs	Uses one of two LINQ queries to return the immediate basis-tree arcs that are connected to the provided parent node. Does not select the arc connecting the parent node and the provided grandparent node
GetNonBasic	Uses a LINQ query to find all non-basic arcs in the provided IList object. It returns an IOrderedEnumerable list of arcs that are used in calculations that only require non-basic arcs, such as determining the reduced cost. Excludes artificial arcs
GetNonBasicWithFLow	Uses a LINQ query to find all non-basic arcs with non-zero flow (such as upper- or lower-bounded arcs). Only used for output-display purposes
GetReversePreorder	Uses a LINQ query to find the basis tree in reverse order. While its functionality could have been created by using a parameterized version of the GetBasis() method, a separate method helps make the purpose more clear in the calling procedures
ResetCycleDirections	Sets the value of all the arc's same cycle-direction flags to true for new cycle calculations
ResetReducedCosts	Sets the value of all the arc's reduced cost values to zero for a new reduced-cost calculation iteration

to as O/R mapping or ORM) is used to read the node and arc class structures from a database and to automatically generate the code for their interactions. Some common methods include getting an object or list of objects from the database as well as inserting, modifying, and deleting object records.

Because the node and arc classes are created as partial classes, the generated database methods could be easily re-generated and stored in separate files if changes are made to the class properties. This eliminated the need to rewrite any methods or to copy and paste code from the generated files into a single class file.

8. **XML Input Files:** The information required by the CTP Solver to represent the network is relatively minimal. A user only needs to supply the following properties for the nodes: id, name, and supply (negative value used for demand). Again, the name property is simply included for human readability and could be omitted from the XML document and require a single line to be commented out in the CTP Solver's code. If memory limitations arise, this would be a good first step to minimize some overhead.

The required arc properties include Tail, Head, Capacity, LowerBound, Cost, and BiDirectional. In addition to the node and arc information, the CTP Solver also requires a configuration element. This element allows the user to set specific values for Big M and the maximum simplex iterations. Enabling these two

Table 12.5 Algorithm methods

Method	Description
GetData	Container method for reading the database or XML file, and initializing the list structures and other variables
GetRecordCounts	Counts the number of nodes and arcs that exist in the network
GetConfigSettings	Reads the configuration options that are set by the user, including the Big M value, maximum simplex iterations, and whether to display the detailed information for each simplex iteration or the cycle-iteration debugging
DisplayOptimal	Builds the optimal solution table and prints the result to the screen
DisplayData	Builds the tables displayed for each simplex iteration, including separate tables for node and arc data, entering and leaving arcs, the basis tree, and cycle arcs. The current total network cost is also included
BuildArcTableHeader	Allows dynamic output table-header creation
GetNodes	Reads the node information from the data and builds the list of nodes as well as the initial basis
GetArcs	Reads the arc information from the data and builds the list of arcs
CalculateSimplex	The main calculation loop of the application. Calls helper methods to calculate the reduced cost, create a cycle, and update the basis for each simplex iteration
CalculateReducedCost	Loops through non-basic arcs to find the entering arc for creating a cycle
CreateCycle	Builds the cycle created by adding the entering arc to the basis, determines the maximum flow change, and chooses and removes the leaving arc from the basis
TraverseBackpath	Determines the basis-tree arc of a given node and depth; helper method used in cycle creation
UpdateBasis	Recursively updates the node preorder values, finds immediate child nodes at a given basis depth, determines arc reflection, and node depths

values to be defined by the users provides some customization to the CTP Solver's capabilities without compromising the application's algorithms with problems such as infinite loops.

In addition to calculation options, the configuration element also allows the user to choose whether to display the full output details for each simplex iteration and/or cycle iteration's debugging information. As shown in the results, disabling this optional information can provide dramatic performance gains on larger networks.

12.5.5 Modified Simplex Algorithm

Recall that the typical steps in the network simplex algorithm [BJS10], pp. 347–348 for the capacitated transshipment problem are as follows:

1. Determine Primal and Dual Solutions
2. Check Optimality

3. Add the Lower-Bounded Arc to the Basis
4. Add the Upper-Bounded Arc to the Basis

Because the CTP Solver was written with object-oriented concepts in mind, it deviates from the standard linear-programming model and uses a modified simplex algorithm to find the optimal solution of a given network; this process is described in the following five steps:

1. Initialize
2. Calculate Reduced Costs
3. Create Cycle
4. Update Basis
5. Repeat Steps 2–4 Until Optimal

In comparison, Steps 1 and 2 are, by and large, performing the same functionality in both the typical algorithm and the CTP Solver's algorithm. The typical Steps 3 and 4 are essentially modified versions of the same step, making slight alterations between the way lower-bounded and upper-bounded arcs are handled and entered through a conditional check determined in Step 2. The CTP Solver effectively combines these two steps into its Step 3 for creating the cycle when either an upper-bounded or lower-bounded arc is added to the basis. The typical Steps 3 and 4 also break into multiple sub-steps that include updating the basis tree. This particular process makes sense because a separate subroutine can be more easily understood as a separate step in the algorithm.

Finally, the CTP Solver's Step 5 is included as a separate algorithmic step for similar reasons as its Step 4, allowing a more simplified description of the process. As with the basis-updating process, this step is also embedded as part of the typical algorithm's Steps 3 and 4.

12.5.5.1 Step 1: Initialize

During initialization, the CTP Solver attempts to read the database or XML file that is chosen by the user. If an XML file is chosen, the application checks to make sure that three tables exist in the file: config, node, and arc.

If the expected number of tables is not present in the XML file, the CTP Solver will stop the execution and print an error message for the user. If the expected number of tables is present in the XML file, the CTP Solver will read the file and populate the lists of nodes and arcs as well as overwrite the default configuration variables, such as Big M and the number of allowed simplex iterations.

1. Set the Root Node

   A root node is simply a starting point for the basis tree (described next). From the root node, the path to all other nodes in the network can be traced. When the CTP Solver reads the node elements from the XML file or database, it creates an artificial node as the default root node with an id value set as the node count. This node is then inserted into the IList of nodes as the last element so that every real

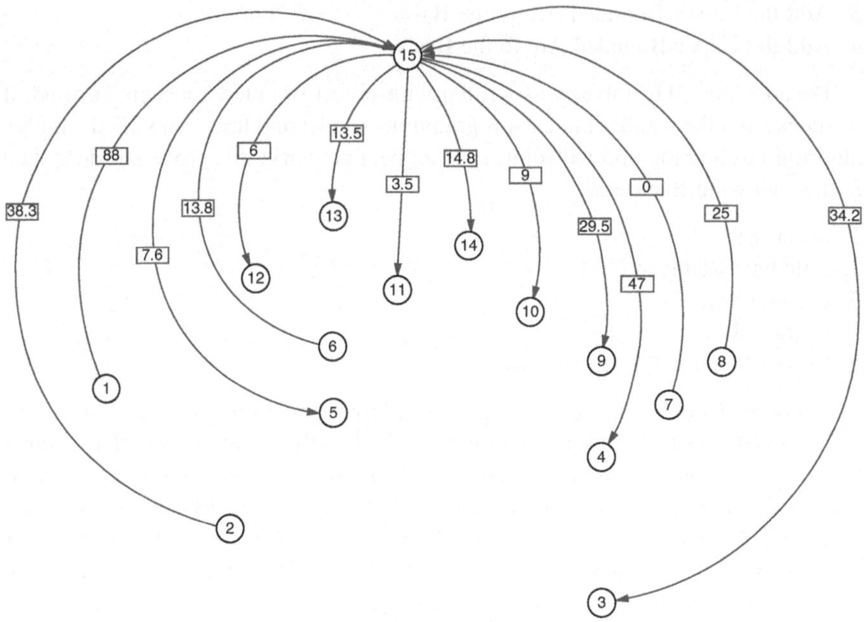

Fig. 12.5 Initial basis tree of the IEEE 14-Bus Test System with arc flows

node can be referenced with its natural id/number (assuming that nodes are ordered numerically starting at 1 in the original network).

2. Create the Basis Tree
   A basis tree is essentially a feasible minimal spanning tree, or a structure where every node is connected by the minimum number of required arcs, and it must be an acyclic-directed graph. An example basis tree is shown in Fig. 12.5.

   As a convenience to the application's user, the CTP Solver automatically creates an initial basis tree to represent a feasible topology of the network, thus removing the need for the user to manually calculate an initial basis and allowing him/her to focus on and only need knowledge about the specific network values. Using the generated artificial node as the root, the CTP Solver creates an artificial arc between every real node and the artificial node, forming the initial basis. Each artificial arc in the initial basis is given a cost for the Big M value that is set by the user in the configuration element, and a flow equal to the absolute value of the node's supply attribute. Using the Big M method as opposed to the Two-Phase method allows the actual arc costs to be used in the first step of the initial basis calculation instead of tracking the original costs as well as reassigning the cost value for every arc in the network.

   In addition to initial cost and flow values, each artificial arc is also given an id, starting with the integer data type maximum value and decrementing as needed. This provides a visual differentiation between real and artificial arcs that is easily seen in the result tables. When determining the head and tail nodes of an artificial arc, the real node's supply value is considered. If the supply is a positive value, the node is considered a supply node with the real node set as the artificial arc's tail and the artificial root node set as the head. If the supply has a negative value, the node is considered a demand node with the real node set as the artificial arc's head and the artificial root node set as the tail. Transshipment nodes are treated in the same manner as supply nodes.

By directing transshipment nodes toward the root node, the initial basis tree is considered strongly feasible. As such, degenerate arcs, or basic arcs with zero unit flow, can be handled without creating an infinite loop that is caused by repeatedly iterating through a sequence of degenerate, basic, feasible solutions that correspond to the same simplex extreme point [CI94], pp. 341–343.

3. Determine the Node Potentials
   In linear-programming terms, the node potentials are equivalent to the dual variables. In algorithmic terms, the node potentials are the summed arc costs along the path of any node back to the root node in the basis tree. Because the initial basis tree essentially consists of a single arc between every real node and the artificial root node, the potential of every node can simply be set by using the user-specified Big M value in the network-configuration settings.

   The Big M value that is defined in terms of the CTP Solver is just large enough to be considered significantly higher than any existing network values for cost or capacity. While it must be a large value, it cannot be too large as to conflict with the limitations of the data types being used (i.e., setting it at the data type's maximum value). Because the CTP Solver uses the Big M value for artificial arc costs, it could potentially be multiplied by itself as many times as there are number of arcs in the network; however, that result is unlike. This means that there must be enough difference between the Big M value and the maximum data-type value to ensure that a very large node potential can be accurately represented and used in the CTP Solver's calculations.

12.5.5.2 Step 2: Calculate Reduced Costs

The reduced cost is the amount that the network's overall total cost could potentially be changed if a given non-basic arc were inserted into the basis. Because the CTP Solver is created with minimization in mind, the best reduced cost belongs to the arc that potentially lowers the total network cost by the greatest amount. The CTP Solver could be utilized for maximization problems by simply using negative cost values. The algorithm still minimizes the optimal solution, but the results can be changed from negative to positive values.

1. Non-Basic Arcs
   By definition, all reduced-arc calculations are carried out on non-basic arcs. These arcs are easily represented in an IList object, allowing fast traversal of just those arcs instead of the entire network.

$R_{ij} = \pi_i - \pi_j - c_{ij}$

Where:
R_{ij} = Reduced Cost of Arc_{ij}
π_i = Tail Node Potential
π_j = Head Node Potential
c_{ij} = Arc Cost

Eq 12.2 Arc reduced-cost calculation

$R'_{ij} = \pi_j - \pi_i - c_{ij}$

Where:

R'_{ij} = Bidirectional Reduced Cost of Arc_{ij}
π_i = Tail Node Potential
π_j = Head Node Potential
c_{ij} = Arc Cost

Eq. 12.3 Bidirectional-arc reduced-cost calculation

The reduced cost of an arc is calculated using Eq. (12.2).

2. Bidirectional Arcs

The CTP Solver handles bidirectional arcs by simply flipping an arc's head and tail nodes for the reduced-cost calculation, as shown in Eq. (12.3).
If an arc is at its lower bound with no flow, this calculation is done immediately after the normal reduced-cost calculation, and the two values are then compared. If the bidirectional reduced cost is better than the original reduced cost, the arc's head and tail nodes are swapped. It is important to reiterate that non-basic upper-bounded arcs and lower-bounded arcs with flow cannot be considered bidirectional because they are already part of the current solution. Allowing these arcs to be treated as bidirectional will often cause net-flow violations, rendering the solution infeasible.

By handling bidirectional arcs in this way, the user does not need to duplicate every instance of an arc when the only difference is the flow direction. This simple implementation is a very important innovation in directed-flow calculations because bidirectional arcs are usually treated as two separate, directed arcs [HSR08], p. 121. The CTP Solver considers the two directions differently even though they are only defined once. In networks consisting of all bidirectional arcs, the CTP Solver effectively halves the size of the required data file, saving both hard-drive space and system memory.

3. Choose the Entering Arc

The preferred reduced-cost value of a non-basic arc could be positive or negative depending on its bounded flow. If the arc has a flow that is equal to its lower bound, a positive reduced cost is desired. If the arc has a flow that is equal to its capacity, a negative reduced cost is desired. The reduced cost of every non-basic arc is compared to the best-available reduced-cost value. When a reduced cost is found to be more attractive, the best-available reduced-cost arc is replaced with the current arc. This process continues until all non-basic arc reduced costs have been determined and the best available reduced cost arc is chosen.

4. Enforce the Lower Bounds

   In order to accommodate an attempted enforcement of the lower-bound flow requirements, any arc with a lower-bound value greater than zero with flow less than the lower bound is given priority. The algorithm forces these arcs into the basis and attempts to push as much flow that is feasible onto them in order to fulfill the lower-bound requirements. With realistic values, this method appears to be sufficient to meet the lower-bound flow requirements. If the CTP Solver finishes its simplex iterations and finds an optimal solution without meeting all the lower-bound flow requirements, the application displays a warning message that a lower-bound flow violation occurred.

   Using the lower bound as an initial flow value is implemented as a possible solution for lower-bound flow enforcement. Unfortunately, determining an elegant process for guaranteed feasibility is not achieved because it is not always clear which artificial arcs could be updated in conjunction with the lower-bounded arc in order to maintain flow balance.

5. Optimality

   The CTP Solver assumes optimality until it encounters an attractive entering arc. If no arcs lower the total network cost when added to the basis tree, the solution is optimal.

12.5.5.3 Step 3: Create Cycle

By definition, the basis tree is a connected graph with no cycles. This means that there is a path between any two nodes but not a path from any node to itself [Knu97], p. 363. When a non-basic arc is added to the basis tree, a cycle is created, and an arc must be removed to preserve the basis tree's acyclic property.

The process of creating a cycle is the most complex step of the CTP Solver's algorithm. If it were a simple shortest-path problem using the arc cost as the arc weight, an algorithm such as Dijkstra's [Dij59] could be used to find an optimal solution. However, the capacitated transshipment problem includes both bounded arcs and directed flow with supply and demand, making it a much more complicated problem (Fig. 12.6).

1. Add Arcs to the Cycle

   Determining the entering arc, or the arc added to the basis tree to form a cycle, is a relatively simple process. Traversing that cycle is a bit more complicated. The algorithm used to create the list of only cycle arcs implements a node-depth concept from [BBG77], following the back path from each entering arc's node to the root node of the basis tree. Using the node depth allows the two back paths to

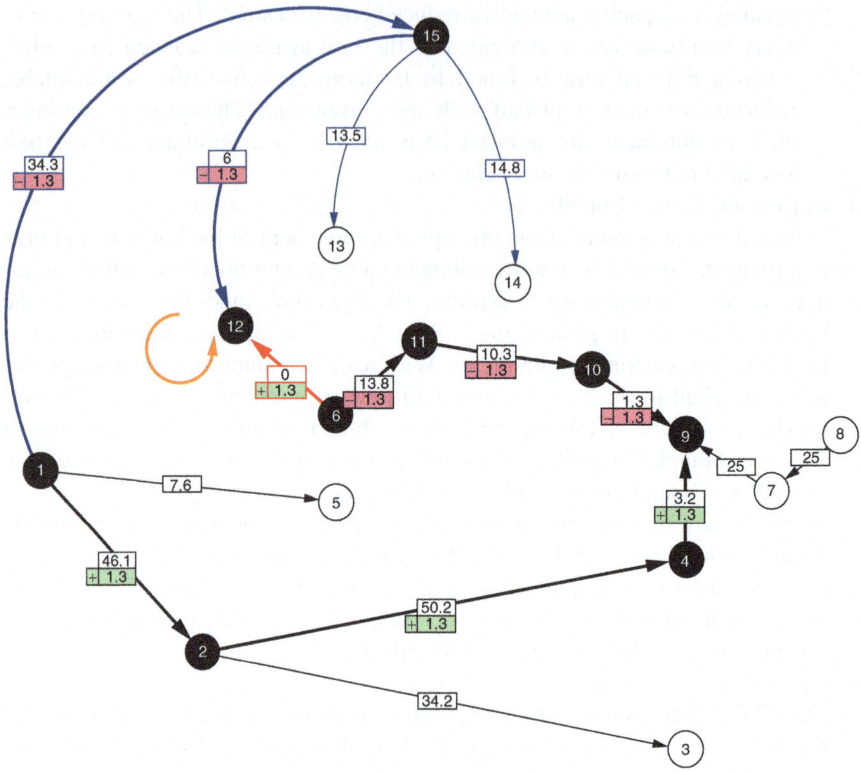

Fig. 12.6 IEEE 14-Bus Test System cycle iteration 13

be traversed in pairs during the same iteration, starting at the deepest node in the cycle (highest depth value) and working back up the basis tree until the two back paths meet at the same parent node, or the root node is reached; either action completes the cycle.

The trick to the CTP Solver's algorithm comes from the need for the entering arc's head and tail nodes to be handled separately in order to account for the correct cycle-direction modifier: a positive one for an arc with flow in the same direction as the entering arc and a negative one for an arc with flow in the opposite direction of the entering arc. By creating a parameterized method for traversing the back path, the same code can be reused with only a few conditional checks to determine the arc's cycle direction. As the node back paths are followed in this manner, the arc connecting each node and its parent is added to the list of cycle arcs if it has not already been added, thus creating the complete cycle.

2. Calculate the Maximum Feasible Flow Change
   As each arc is added to the cycle, its maximum feasible flow change is calculated based on the arc's direction in relation to the cycle created by the entering arc. This value, represented by the Greek letter theta, is the largest number of flow

units that could be added or subtracted from a same- or opposite-cycle direction arc, respectively, without violating the arc's flow-capacity or lower-bound requirement.

Using the theta value, the CTP Solver's algorithm ensures that each simplex iteration moves the current basis tree as close to the optimal solution as is feasibly possible. For same-cycle direction arcs, the theta value is simply the difference between the arc's capacity and its current unit flow. For opposite-cycle direction arcs, the theta value is calculated as the difference between the arc's current unit flow and its lower bound.

3. Choose the Leaving Arc
   Once an arc's theta value has been determined, it is compared against the cycle's current minimum theta value. To maintain feasibility when an arc is removed from the cycle, the smallest theta value from all the cycle arcs must be used to ensure that no capacity or lower-bound violations occur.

   The cycle's minimum theta value can only be changed if the current arc's theta value is strictly less than the cycle's overall minimum or if the arc is artificial. These two theta-updating conditions prevent infinite cycles due to degeneracy and force artificial arcs from the basis, respectively.
4. Update the Cycle Flows
   Once the leaving arc has been determined, the cycle is iterated a final time in order to add or remove theta units of flow to its arcs. Using an arc's direction property, flow is added to same-direction cycle arcs and subtracted from opposite-direction cycle arcs. By following the cycle direction, the solution's feasibility is ensured because arc limits are not capable of being violated by adding or subtracting too many flow units.
5. Degeneracy
   Recall, from the initial basis creation, that a degenerate arc is one with a unit flow of zero. Degenerate arcs can cause infinite loops and must be handled properly to avoid such problematic outcomes. The initial basis is created to be strongly feasible, and the CTP Solver needs to maintain that status.

   Using the node-depth method described for the cycle creation, the lowest (deepest) degenerate arc can be chosen to leave the basis, preserving a strongly feasible basis [BBG77]. This is accomplished with the algorithm's tie-breaking conditional check that occurs when determining the cycle's minimum theta value. Because the algorithm starts at the deepest cycle arc, a simple comparison can be made between the current theta value and the cycle's minimum value and only change the value of theta if the former is less than the latter, thus always choosing the deepest cycle arc.

12.5.5.4 Step 4: Update Basis

Updating the basis involves a recursive method, or a method that calls itself, starting from the root node as the top of the basis tree and working down one node level at a time until all nodes and arcs have been updated. Possible errors could result in allocating the system's memory [Som11] during the recursive process; however, the node and arc structures used in the method are already stored in the memory using the IList structures. If memory allocation is an issue, it would likely occur before the recursive process begins.

A possible optimization, discussed later, would be to only update the cycle nodes and arcs instead of the entire basis tree. Using recursion through the full-basis tree is chosen due to its simplification of the algorithmic process, essentially implementing an easily comprehended depth-first search [RN10], p. 85. The search space for each iteration is the size of a spanning tree, or one less than the number of nodes in the network [HSR08], p. 236. During each recursive iteration of the basis update method, the CTP Solver uses a LINQ query to find all of the current node's child nodes. Then for each child node, the method calls itself to find that node's child nodes. This process repeats until the entire basis tree has been traversed, and the node potential and depth values have all been updated. Once the basis tree has been updated, it is ready for use with the next simplex iteration unless it is already optimal.

12.5.5.5 Step 5: Repeat Steps 2–4 Until Optimal

The CTP Solver assumes optimality until the non-basic-arc reduced-cost values are calculated in Step 2. If adding a non-basic arc to the basis lowers the overall total cost of the network solution, the optimality flag is set to false, and the CTP Solver executes another simplex iteration, continuing through Steps 3 and 4 (Fig. 12.7).

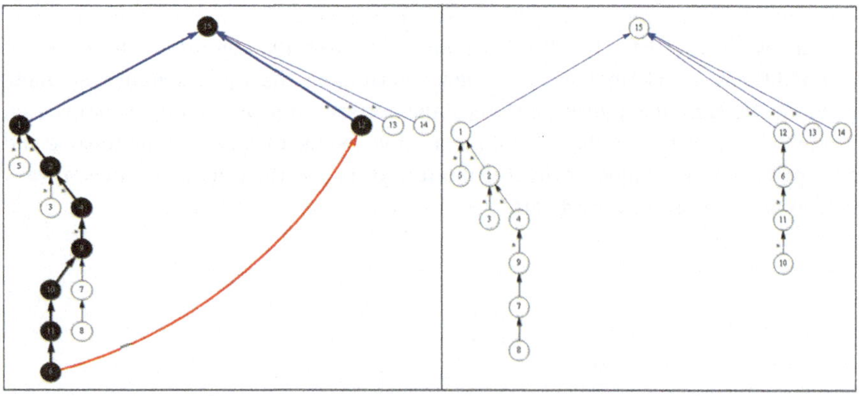

Fig. 12.7 IEEE 14-Bus Test System basis update 13

Once the reduced-cost calculation step determines that there are no non-basic arcs that should become entering arcs, the process is complete, and the solution is optimal.

12.5.6 Output

The CTP Solver provides the user with all available information about the resulting network in addition to the step-by-step simplex iterations and the cycle debugging log. The user is also shown a link to the original network file and the optimal solution that are available for download in either XML or .CSV format.

12.5.6.1 Miscellaneous Information

Various information about the results is displayed to the user before any other data. First, there are links to the network files, including the original network (if the source was an XML file) as well as downloadable .CSV and XML files for the optimal solution. After the network file links, the optimal network cost, number of simplex iterations, node count, basic-arc count, non-basic arcs with flow count, and execution time are all displayed.

Utilized in conjunction with the optimal network table, the user is able to quickly understand the results of the CTP Solver and to download the information for analysis or importing into other systems. A screenshot of the miscellaneous information for the IEEE 14-bus test system using distance as a cost measure is shown in Fig. 12.8.

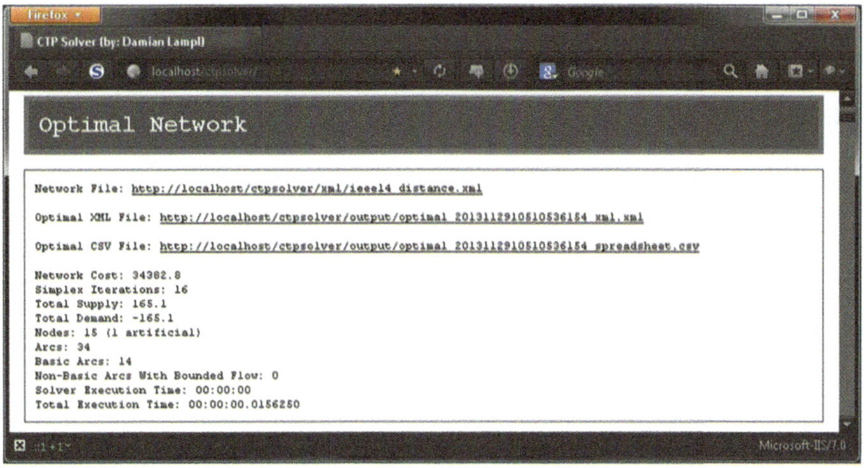

Fig. 12.8 CTP solver's optimal-network miscellaneous information (screenshot and light CSS)

12.5.6.2 Optimal Solution

After the miscellaneous result information, the optimal network's table is displayed on the results page, showing the optimal network's topology via the list of arcs. Because the node information is not required to reconstruct the optimal network, the list of nodes is omitted.

The table showing the optimal solution includes each arc's id, tail node, head node, cost, capacity, lower bound, flow, capacity minus flow, reduced cost, basis order, and whether the arc is basic or an arc with bounded flow (denoted as "non-basic"). All non-basic arcs without bounded flow are simply displayed with a hyphen for the basis value. Artificial arcs are included in the final optimal display table; however, they are visually separated from the real-network arcs and are not included in the XML or .CSV exports. A screenshot of the optimal solution for the IEEE 14-bus test system using distance as a cost measure is shown in Fig. 12.9.

12.5.6.3 Simplex Iterations

The information for each simplex iteration can be shown if the user so chooses in the configuration options using the "showSimplexIterations" attribute. When shown, every iteration is given a separate, expandable block of information that details a snapshot of the network. In addition to the same information being displayed for the optimal network's list of arcs, each cycle, the entering and leaving arcs, and an individual table for the basis tree are shown.

During the debugging process, showing the data for each simplex iteration can be very useful as a way to step through the algorithm's process to follow every decision made for verification purposes. The first iteration is the network as it is provided to the CTP Solver. Successive iterations show the evolving network as the algorithm progresses.

12.5.6.4 Debugging Log

The debugging log includes the cycle created by each simplex iteration. As with the simplex iterations, the cycle's debugging log can be toggled by the user in the network file's configuration element using the "showDebuggingLog" attribute. Each cycle begins with its entering arc and shows each cycle arc's direction in relation to that entering arc, along with its maximum feasible flow change. Just like the simplex iterations, the cycle's debugging log can be valuable to trace the CTP Solver's algorithmic process for verification.

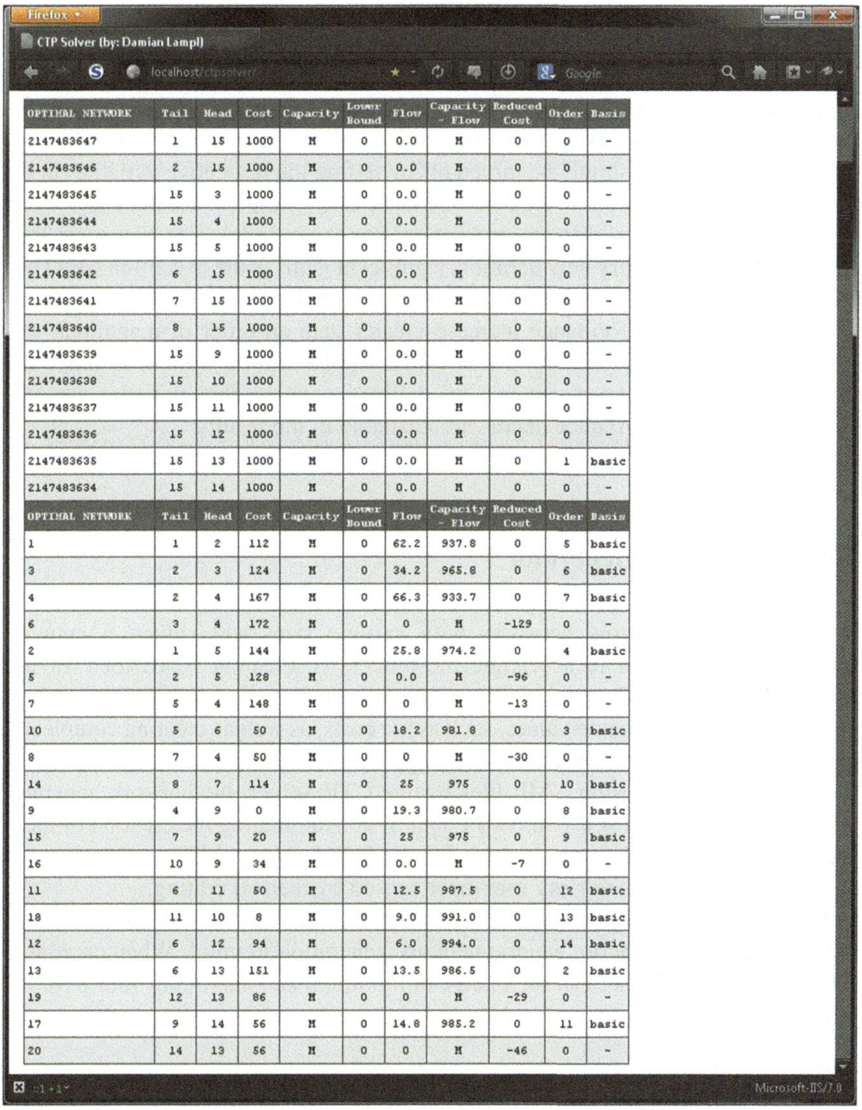

OPTIMAL NETWORK	Tail	Head	Cost	Capacity	Lower Bound	Flow	Capacity - Flow	Reduced Cost	Order	Basis
2147483647	1	15	1000	M	0	0.0	M	0	0	-
2147483646	2	15	1000	M	0	0.0	M	0	0	-
2147483645	15	3	1000	M	0	0.0	M	0	0	-
2147483644	15	4	1000	M	0	0.0	M	0	0	-
2147483643	15	5	1000	M	0	0.0	M	0	0	-
2147483642	6	15	1000	M	0	0.0	M	0	0	-
2147483641	7	15	1000	M	0	0	M	0	0	-
2147483640	8	15	1000	M	0	0	M	0	0	-
2147483639	15	9	1000	M	0	0.0	M	0	0	-
2147483638	15	10	1000	M	0	0.0	M	0	0	-
2147483637	15	11	1000	M	0	0.0	M	0	0	-
2147483636	15	12	1000	M	0	0.0	M	0	0	-
2147483635	15	13	1000	M	0	0.0	M	0	1	basic
2147483634	15	14	1000	M	0	0.0	M	0	0	-
OPTIMAL NETWORK	Tail	Head	Cost	Capacity	Lower Bound	Flow	Capacity - Flow	Reduced Cost	Order	Basis
1	1	2	112	M	0	62.2	937.8	0	5	basic
3	2	3	124	M	0	34.2	965.8	0	6	basic
4	2	4	167	M	0	66.3	933.7	0	7	basic
6	3	4	172	M	0	0	M	-129	0	-
2	1	5	144	M	0	25.8	974.2	0	4	basic
5	2	5	128	M	0	0.0	M	-96	0	-
7	5	4	148	M	0	0	M	-13	0	-
10	5	6	50	M	0	18.2	981.8	0	3	basic
8	7	4	50	M	0	0	M	-30	0	-
14	8	7	114	M	0	25	975	0	10	basic
9	4	9	0	M	0	19.3	980.7	0	8	basic
15	7	9	20	M	0	25	975	0	9	basic
16	10	9	34	M	0	0.0	M	-7	0	-
11	6	11	50	M	0	12.5	987.5	0	12	basic
18	11	10	8	M	0	9.0	991.0	0	13	basic
12	6	12	94	M	0	6.0	994.0	0	14	basic
13	6	13	151	M	0	13.5	986.5	0	2	basic
19	12	13	86	M	0	0	M	-29	0	-
17	9	14	56	M	0	14.8	985.2	0	11	basic
20	14	13	56	M	0	0	M	-46	0	-

Fig. 12.9 CTP solver's arc information for an optimal network (screenshot and light CSS)

12.5.7 *Limitations and Modifications*

Not all networks are guaranteed to have a feasible solution. The CTP Solver handles these networks by providing warning messages prior to displaying the best possible solution that the application obtained. These warning messages alert the user to infeasibilities, such as lower-bound flow violations, artificial arcs unable to be removed from the basis, and net-flow violations on any nodes.

12.5.7.1 Performance Gains

The CTP Solver displays a lot of information. However, some users may not be interested in the output generated for every simplex iteration or the debugging log that shows each cycle. With these users in mind, the debugging and individual simplex iterations can simply be turned off by setting the respective variables in the network file's configuration element. Because each simplex iteration and cycle traversal generates data that are proportional to the network size that must then be displayed during every new iteration, significant gains in the execution speed can be achieved by choosing to hide this information for larger networks.

For example, a performance increase greater than an order of magnitude was observed for a test network with just 30 nodes and 55 arcs by hiding the individual simplex iterations. That performance was doubled when the debugging log was also hidden. The full performance details are available in the results.

12.5.8 Network Generator

In order to test multiple networks with varying sizes and values, a network generator is developed to accelerate the process of creating XML files for use with the CTP Solver. The network generator can read the existing IEEE test-system text files and generate some values, such as arc costs, as well as creating completely random networks with user-defined topology values and limits.

In addition to exporting XML files for the CTP Solver, the network generator also exports data files for use in AMPL and SAS. Automating the creation of these additional files makes comparisons among the CTP Solver, AMPL, and SAS much easier while also removing any user error caused by manual editing.

In addition to reading the standard IEEE test-system files, the network generator is built with pseudorandom generic-network generation in mind. When creating a generic network, the user can set a range of minimum and maximum values for the following components:

1. Node Supply
2. Arc Capacity
3. Arc Lower Bound
4. Arc Cost

In addition, the user can set values for the total network supply and a lower-bound frequency threshold, defined as an integer value from 1 to 100 essentially acting as a percentage for approximately how often the user would like a lower-bound value to occur for network arcs.

12.6 Software Comparisons

The CTP Solver uses a customized simplex algorithm that is implemented in ASP. NET C#. The results of multiple test networks are compared with two optimization software programs: AMPL and SAS.

12.6.1 AMPL

AMPL is a comprehensive and powerful algebraic modeling language for linear and nonlinear optimization problems; AMPL was developed at Bell Laboratories [FGK03]. One of the primary benefits of AMPL is its separation between the model and data files, allowing the user to utilize the same model for multiple datasets. The user is expected to learn AMPL's syntax to create his/her own models for specific applications. AMPL also allows the user to choose from many different custom-solution solvers. The solver chosen for the test networks was LPSOLVE, an open-source simplex solver.

The AMPL version used on the test networks was AMPL Student Version 20100715 (MS VC++ 6.0). LPSOLVE solver version 4.0.1.0 was used because it allowed the highest number of variables and constraints with the student version of AMPL. Due to these software limitations, the IEEE 300-Bus Test System could not be solved using AMPL's student version because it had too many variables and constraints.

12.6.2 SAS

SAS is a collection of software solutions that are used to solve complex business problems based on three key capabilities: information management, analytics, and business intelligence [SAS].

As with AMPL, SAS is a powerful software tool with the ability to separate a problem model from its data. SAS requires its users to learn its programming-language syntax in order to create their own models. The built-in SAS method used for the test networks was the NETFLOW procedure. Unlike AMPL's LPSOLVE solver and the CTP Solver, the NETFLOW procedure uses the interior point algorithm [SAP] instead of the simplex algorithm. It also uses the "good-path" method described in "Algorithms for Networking Programming" by J. Kennington and R. V. Helgason [SAS10]. Version 9.3 of the X64_VSPRO platform of the SAS software was used for the test networks.

12.6.3 CTP Solver

Due to the design goals of the CTP Solver's implementation, it offers key benefits to its users that are not provided by AMPL or SAS. Most of these benefits are directly related to the CTP Solver's ease-of-use design goal that focuses on simplicity for the users.

1. Modeling: While incredibly robust and capable solutions, both AMPL and SAS require the users to understand how to model the problems in order to understand and utilize the solutions. This allows many more problem types to be solved, but the learning curve may be too steep for most users because each software application has its own syntax. Because the CTP Solver abstracts the user from the modeling process, it can simply be utilized with properly formatted XML files or a database.
2. Software Installation: The installation process for SAS, in particular, can be an overwhelming experience for typical users, requiring both the SAS-software as well as Java-runtime dependencies. It includes different business analytics, intelligence, and information-management solutions, making it an extremely complex process before using the software. AMPL does not need installation, but it does require downloading and extraction, as well as user knowledge of the program's file structure in order to find the data and model files. Because the CTP Solver is a web application, a user simply needs a browser to access and utilize it, making it more accessible than either AMPL or SAS.
3. Output: Both AMPL and SAS produce simplified results by default, with SAS more closely resembling the CTP Solver's default table output. However, the CTP Solver exports its results to XML and .CSV data files by default, or directly to the database if the input network is a database source. Customized output requires more user effort with both AMPL and SAS than with the CTP Solver.

12.6.4 Displayed Results' Comparison

All three applications used to test the networks display information to the user in their own way. The primary aspects of particular interest are displaying the optimal solution value, along with key diagnostic information regarding the solver's performance, and the topology of the resulting optimal network including flows.

1. Optimal Display and Diagnostic Comparison
   Both SAS and the CTP Solver do a relatively good job providing the user with important, detailed information in an easy manner. AMPL requires a bit more user effort to find some of the relevant data.
   The CTP Solver displays this important information immediately before the optimal network topology, allowing the user to quickly determine key aspects of the results. SAS also displays its information in a way that is easy for the user to access by utilizing its log-output window. When using AMPL, the user must

specifically print the diagnostics that he/she is interested in viewing. This requirement subjectively makes the diagnostics' display in AMPL a bit more cumbersome than SAS or the CTP Solver because the user must read through the documentation and become familiar with the relevant variables and how they are used in AMPL.

2. Optimal Network Comparison
   In addition to important information about the results, all three applications provide a representation of the optimal network with flows along the arcs.
   The CTP Solver separates itself from AMPL and SAS with a few key features, including exporting the resulting network to both XML and CSV for easy use in other applications and portability among different systems. Another nice feature of the CTP Solver is the inclusion of repeating table headers every 20 rows after the artificial arcs, allowing the user to easily see which information is in each table cell at a glance instead of scrolling all the way back to the top as SAS requires. The CTP Solver also differentiates between basic and non-basic arcs with flow, providing the user with more detailed information when bounded arcs are included in the solution.

12.6.5 Accuracy Summary

Accuracy was determined by comparing the resulting optimal network flow computed by the CTP Solver to the optimal network flow computed separately by both AMPL and SAS for the same network.

The CTP Solver, AMPL, and SAS each employ optimal algorithms, so it should be expected that they all obtain the same optimal result for the test networks. This was the case with the tests performed, and because many optimal solvers currently exist, new contributions should focus on improving performance and ease of use. Because ease of use is inherently subjective, the software performance, in terms of speed, should be considered the best comparison measure for the three solutions.

12.6.6 Performance Summary

On smaller networks, up to and including the IEEE 118-Bus Test System, there was little difference among AMPL, SAS, and the CTP Solver. The computation time was low enough that the measurement precision could be questioned due to the way processing time is essentially estimated using the CTP Solver's system clock. The true performance comparison came from the larger, randomly generated test networks. Unfortunately, the student-license version of AMPL was only able to test up to the IEEE 118-Bus Test System, so it was primarily a contest between the CTP Solver and SAS.

Sadly, the CTP Solver was destroyed by SAS on larger networks; it was not even close. Somehow, and very surprisingly, both the AMPL LPSOLVE solver and the SAS "netflow" procedure maintained a very consistent execution time throughout all tests, even when the network size increased. The expectation was for the execution time to become progressively higher as the node and arc counts grew, as was the case with the CTP Solver.

During the CTP Solver's implementation, the hope was that LINQ queries would be fast enough to overcome the recursive traversal through the entire basis tree for each simplex iteration. Unfortunately, this likely contributed to its poor performance on larger networks. While this outcome was extremely disappointing, the CTP Solver is not a wasted effort. Improvements can clearly be made to its modified simplex process, and some possibilities are outlined in the Conclusion. Its handling of bidirectional arcs is also an encouraging innovation that could be utilized in other systems.

If the network size were small enough, the CTP Solver might be an optimal self-healing method for the smart grid. Realistically, the CTP Solver, in its current state, is simply too slow to be considered a viable solution. Interestingly, a couple of the CTP Solver's performance improvements were implemented before comparing the results to SAS.

12.6.6.1 The CTP Solver's Output-Performance Improvements

For the larger test-network comparisons, displaying each simplex iteration and cycle traversal was unnecessary bloat. The configuration allows each display option to be shown or hidden, allowing the user to decide whether the CTP Solver should output the information.

Choosing to only display the final, optimal network provides significant speed improvements. For each of the following tests, ten runs were made for each average solve time along with the final results from the application comparison runs, providing an approximate general-performance result.

The CTP Solver can display the entire network topology at every simplex iteration, allowing the user to step through the results and to follow the solver's decisions. This can be beneficial when manually calculating solutions, such as verifying student results for assignments in an academic course. However, for large networks, manual solution calculations are simply infeasible, which is the entire purpose of software such as the CTP Solver. As such, the output for these individual simplex iterations can be set to not display, greatly improving the CTP Solver's execution-time performance. Tables 12.6–12.8 list the system's performance when different display options are hidden during program execution.

Another attempt to improve the CTP Solver's performance was altering the pricing or reduced-cost calculations (Step 2). In order to choose the best candidate arc to enter the basis, this process iterates through every non-basic arc. While the calculations are fast with the current machines, the number of comparisons can

Table 12.6 The CTP Solver's performance improvement: hiding simplex iterations

Average solve time (s)	Display all	Hide simplex iterations	Improvement
IEEE 30-Bus	0.071875	0.00625	**11.5**× **(1050%)**
IEEE 57-Bus	0.290625	0.0234375	**12.4**× **(1140%)**
IEEE 118-Bus	2.529513889	0.168402778	**15.02**× **(1402%)**

Table 12.7 The CTP Solver's performance improvement: hiding the cycle debugging log

Average solve time (s)	Display all	Hide cycle debugging log	Improvement
IEEE 30-Bus	0.071875	0.059375	**1.21**× **(21%)**
IEEE 57-Bus	0.290625	0.28125	**1.03**× **(3%)**
IEEE 118-Bus	2.529513889	2.310763889	**1.09**× **(9%)**

Table 12.8 The CTP Solver's performance improvement: hiding the cycle debugging log and simplex iterations

Average solve time (s)	Display all	Hide all	Improvement
IEEE 30-Bus	0.071875	0.00625	11.5× (1050%)
IEEE 57-Bus	0.290625	0.021875	13.29× (1229%)
IEEE 118-Bus	2.529513889	0.16875	14.99× (1399%)

potentially decrease by three orders of magnitude in networks with thousands of arcs.

12.6.7 Testing Environment and Setup

In order to maintain consistency for each of the three applications being compared, the test networks were solved on the same machine. All tests were performed on an Intel Core 2 Quad 2.67 GHz processor with 4 GB DDR2 800 RAM running on Windows Vista x64. Both AMPL and SAS are standalone software applications. The CTP Solver requires a web server, so a local virtual directory was created for it using IIS, running the .NET 4.0 framework.

12.6.8 Test Network Results

All three software applications were compared using the IEEE 14-Bus Test System, IEEE 30-Bus Test System, IEEE 57-Bus Test System, and IEEE 118-Bus Test System. From there, the student-license version of AMPL was unable to calculate the results due to variable and constraint limits, so only the CTP Solver and SAS were used to compare the IEEE 300-Bus Test System and the Custom 400- and 500-Node Systems. For each network, a series of ten consecutive runs were

Table 12.9 IEEE 30-Bus performance results

Solve time (s)	AMPL (35 iterations)	SAS (37 iterations)	CTP Solver (31 iterations)
Test 1	0.015625	0.23	0.03125
Test 2	0	0.24	0.015625
Test 3	0.015625	0.25	0
Test 4	0	0.21	0
Test 5	0	0.26	0.015625
Test 6	0.015625	0.2	0
Test 7	0.015625	0.23	0
Test 8	0.015625	0.23	0
Test 9	0.015625	0.21	0
Test 10	0	0.18	0
Average	0.009375	0.224	**0.00625**

executed for each solver, and an average of these runs was taken as the solver's general performance time. While this sample was admittedly a small, the intent was to simply make a pedestrian comparison of the three solutions. The detailed test results from the IEEE 30-bus system case are given below.

12.6.8.1 IEEE 30-Bus Results: Random Cost

The CTP Solver was on top, again, in this small network, but AMPL could have had a better overall performance with a larger test sample size.

1. Accuracy
   All three applications, again, arrived at the same optimal solution value: 23,692.6396. The CTP Solver had the fewest iterations at 31, with AMPL and SAS following and requiring 35 and 37 iterations, respectively.
2. Performance (Table 12.9)
3. Performance Graphs (Figs. 12.10 and 12.11)

12.7 Results and Conclusion

In small-network tests, the CTP Solver excelled over SAS and also had slightly better performance than AMPL. However, the larger the networks became, the worse the CTP Solver performed.

SAS kept its growth pattern fairly linear as networks with more nodes and arcs were calculated. Unfortunately, the CTP Solver's growth pattern appeared to be exponential. This was likely due to its traversal of the entire basis tree for each

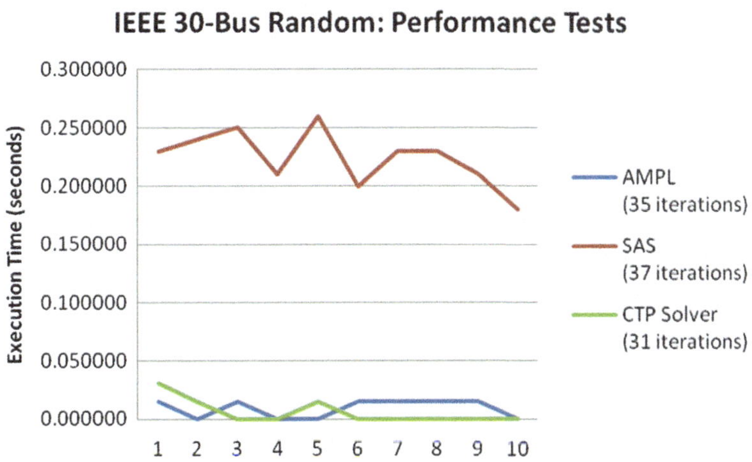

Fig. 12.10 IEEE 30-Bus performance graph

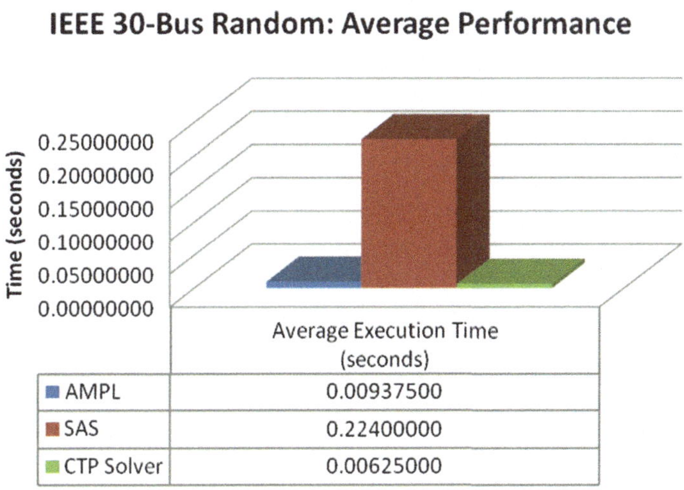

Fig. 12.11 IEEE 30-Bus average performance graph

simplex iteration, making the time complexity approximately $O(n^2)$, where "n" is the number of nodes in the network.

It was hoped that using LINQ queries would compensate, but clearly the way the algorithms were implemented, that was not the case, proving that math trumps faith. An ongoing effort is attempting to resolve this issue while staying true to the use of LINQ queries, but it might come down to reverting to traditional linear-programming techniques as the best approach.

Figure 12.12 shows the average performance results starting from the IEEE 14-Bus System and progressing to the Random 500-Node test network. Only SAS

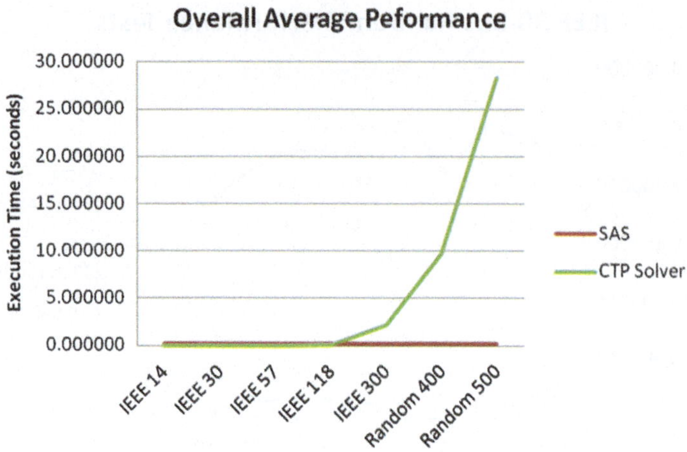

Fig. 12.12 Overall average performance

and the CTP Solver were graphed due to the fact AMPL was only tested on the four smallest networks. The graph clearly shows the exponential growth pattern of the CTP Solver and the linear nature of SAS.

One silver lining for the CTP Solver's performance was its consistency in requiring the fewest iterations to calculate its results. This leads to the possibility of it potentially being much faster in a parallelized implementation.

Chapter 13
Decomposition of Microgrids in Large-Scale Electric Test Beds for Economic Dispatch Optimization

13.1 Introduction

Microgrid decomposition (partitioning, splitting, and clustering) or otherwise determining community structures within a power transmission network is important to optimal management of the transmission system. Power transmission system decomposition is not itself a novel concept. Similar concepts have been utilized dating back to the 1950s for various reasons [kro63]. There are very limited existing literatures that use clustering for grid decomposition.

The earliest work involving grid decomposition focused on the development methods for breaking large systems into smaller subsystems in order to make complex analysis or computations more simple [kro63]. More recent works have focused on the identification of power network zones within a grid [GOB16], spectral clustering of power grids [Ruo13], and assessment of grid reliability based on topological metrics [GTC07]. In [Ruo13], hierarchical spectral clustering methods were used for power grid decomposition, and [GOB16] uses electrical distance quantification as a parameter for dividing a bus system into microgrid-like zones. To our knowledge, there is no paper that has considered an approach similar to the approach proposed in this chapter. Metrics of betweenness centrality (BC) and two-stage clustering bring novelty to grid decomposition. The importance of efficient grid decomposition and microgrid utilization is becoming especially important in considering emerging technologies related to smart grid. Optimal grid decomposition will play an important role in uncertainty quantification, contingency planning, resource allocation, optimal power flow, cascading failure protection, incorporation of renewable power sources, and incorporation of renewable power sharing components of smart grid [SFN14].

P. Ranganathan, K.E. Nygard, *Distributed Linear Programming Models in a Smart Grid*, Power Electronics and Power Systems, DOI 10.1007/978-3-319-52617-1_13

13.2 Background on Smart Grid Modeling Using Graph Theory

13.2.1 Graph Theory and Network Community

A graph is a mathematical structure that represents pairwise relationships or connections between objects [BHP09]. A graph is a set of vertices (nodes) that are connected edges (lines). Because of the connectedness between objects in a power transmission grid, these grids can be intuitively represented as a network graph. Formally, notation for a graph is given as a pair of sets $G = (V, E)$, where G is the graph, V is the set of vertices, and E is the set of edges, formed by the pairs of vertices. A vertex set is a concatenated list of the name of each vertex in a graph and is denoted by $V(G)$. An edge list is a concatenated list of connected vertices in a graph and is denoted by $E(G)$. As an example, Fig. 13.1 shows a graph with $V(G) = \{1, 2, 3, 4, 5\}$ and $E(G) = \{1-2, 2-4, 2-5, 3-4, 3-5, 4-5\}$.

Figure 13.1 displays a simple un-directed graph. An un-directed graph is a graph where the edges of the graph are bidirectional [BHP09]. For the purposes of this work, bus system graphs are considered to be un-directed graphs. The following subsections describe criteria related to graph theory that can be used to determine grid decomposition structures.

Weighted Graphs: The raw graph topology of a given graph does not provide or display any functional information about the actual system the graph represents. In terms of a power grid system, there is information about the buses and transmission lines within the grid that must be considered in order to adequately model the system. This can be accomplished via the use of vertex and edge weights. Weights are simply a numeric value assigned to graph objects to convey some functional information about the graph or specific graph object. A common example of an edge weight is assigning a numerical value to a particular edge corresponding to the length of that edge. For this chapter, the notation for an edge weight is *W_(m,n)*, where m and n are vertices in $V(G)$ such that *W_(m,n)* is the weighted value for the edge that connects bus m to bus n.

In this chapter, four metrics have been considered for edge weights in a smart grid power transmission system. These metrics are topological, admittance,

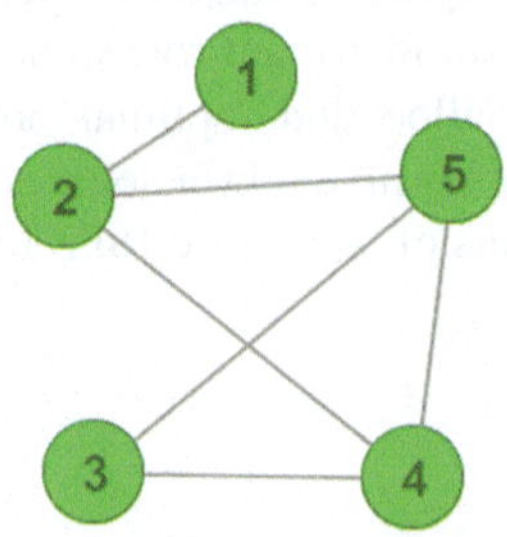

Fig. 13.1 Simple graph where $V(G) = \{1,2,3,4,5\}$ and $E(G) = \{1–2,2–4,2–5,3–4,3–5,4–5\}$

impedance, and line length weights. Topological weight assumes that $W_{m,n} = 1$ $\forall m, n \in E(G)$. The topological weighting metric captures the trivial topological connections of the network graph and displays no bias toward certain network objects. Admittance-based edge weights are determined based upon a calculation of transmission line admittance. For this metric admittance weight is given by: $W_{m,n} = \frac{1}{|R_{m,n}+jX_{m,n}|}$, where $R_{m,n}$ is the resistance of the transmission line connecting bus m to bus n and $X_{m,n}$ is the reactance of the line. Impedance weight is the inverse of the admittance weight. Length-based weighting assigns $W_{m,n}$ equal to the length of the transmission line. For this work, IEEE 57, 118, and 300 bus test systems were used and approximations of transmission line length were calculated according to the method outlined in [Fre77]. This method first converts the per unit reactance value to the actual value using an assumed $S_{\text{base}} = 100$ MVA and $V_{\text{base}} = 135$ kV. The length of the line is then calculated assuming a conversion factor of 0.7 Ω/mile.

These weights are static weights, in that they are constant for a given power system. Other works have considered similar static metrics as well as dynamic metrics including power flow [Ruo13]. The static edge weights can be interpreted such that strongly connected vertices are more likely to be clustered together. Topological weights represent a pure connectivity of the network. Admittance/impedance weights represent and reveal the internal electrical structure based on impedance or electrical distance of the network [GOB16], [Ruo13].

Shortest Path and Betweenness Centrality: Betweenness centrality [Dan60] is an index that quantifies a vertex or edge's centrality in a network. In order to understand betweenness centrality, the graph theory concept of shortest paths needs to be understood. The shortest path problem [JMF99] is a common concept in a study of graph theory. The problem is defined by the task of finding the path between two given vertices in a graph such that the sum of the edge weights of the constituent edges of the path is minimized. A path in an un-directed graph is denoted by $P = \{v_m, v_1, v_2 \ldots v_n\}$, where P is the path and $v_m : v_n$ are vertices in graph G that are contained in the path from v_m to v_n.

The formal definition of betweenness centrality is "the number of shortest paths from all vertices in a graph to all other vertices in the graph that pass through a particular object."[Dan60], [GN02], [GN03] Betweenness centrality can be calculated for vertices or edges. Either of these calculations indicates how central, connectively important, or "highly traveled" a particular edge or vertex is within a graph. This metric is of importance for a smart grid transmission system due to the ability to quantify vertices or edges that are of high connective importance to the network. Buses and/or transmission lines with relatively high betweenness centrality may be more likely to cause cascading problems in the event of a failure to that particular bus or line.

13.2.2 Graph Clustering and Application to Grid Decomposition

Graph clustering is an application of clustering algorithms to perform grouping of nodes in graphs and is related to clustering techniques in the field of data mining. General cluster analysis is the task of grouping a set of objects or data points in such a way that objects in a particular group are more similar to each other than to objects contained in other groups or subsets [New06]. The purpose of clustering is to get an improved understanding of the input group or dataset. In this work methods of clustering are applied to power transmission systems to form power zone community structures that for intents of analysis are designated as microgrids. Graph clustering is a term with several aliases depending upon the application. In general, graph clustering, network community detection, graph partitioning, and graph decomposition are different aliases by which similar processes are occurring. These aliases all mean to discover community relationships between nodes within a graph. These communities are characterized by relatively dense interconnections, but relatively sparse connections between groups. Graph clustering algorithms are designed to identify and quantify where these community structures exist within a graph.

1. **Betweenness Centrality Graph Clustering (BCGC)**: A number of algorithms exist that perform graph clustering based on calculations of the betweenness centrality of graph objects. BCGC algorithms make use of the betweenness centrality as a method of determining the likelihood that an edge is between community structures in a graph. A notable algorithm for betweenness centrality clustering is the Girvan–Newman (GN) algorithm [GN02], [RAK07].

   The GN algorithm detects communities or clusters within a graph by iteratively removing edges from the original network graph. After the removal of edges, the remaining connected components of the network graph are the communities. The GN algorithm removes edges based upon the betweenness index of each edge. Removing edges with high betweenness is a method of separating community structures within a graph from one another. The steps of the GN are as follows:

   (a) The betweenness of all edges within a graph were calculated.
   (b) The edge with the highest betweenness is removed.
   (c) The betweenness of all edges affected by the removal of this edge are then recalculated
   (d) Repeat starting from step 2 until a desired cutoff has been obtained[GN02], [RAK07].

   The stopping point or cutoff of the algorithm can be determined in terms of iterations, a desired betweenness, an optimality of graph modularity [SFO13], when a desired number of clusters has been formed, or when there are no more edges to be removed. The iGraph package in R software has a ready implementation function for this algorithm which was utilized for this work.

The algorithm is related to agglomerative hierarchical clustering algorithms due to the step-by-step way the algorithm decomposes the original graph. This step-by-step decomposition can be viewed as a dendogram, or a hierarchical tree. An example of a hierarchical dendogram is shown in Fig. 13.2b. Figure 13.2a shows an example graph that corresponds to the dendogram in Fig. 13.2b. The colored lines that enclose vertices and the dendogram branches represent the clusters formed in this graph by the GN algorithm. The *y* axis displays the betweenness centrality metric which can be used as a cutoff in the GN algorithm. The *x* axis of the "tree" in 2B represent vertices in the graph. The dendogram branches show the order in which the vertices form community structure per the agglomerative nature of the GN algorithm. At the top of the dendogram, all the

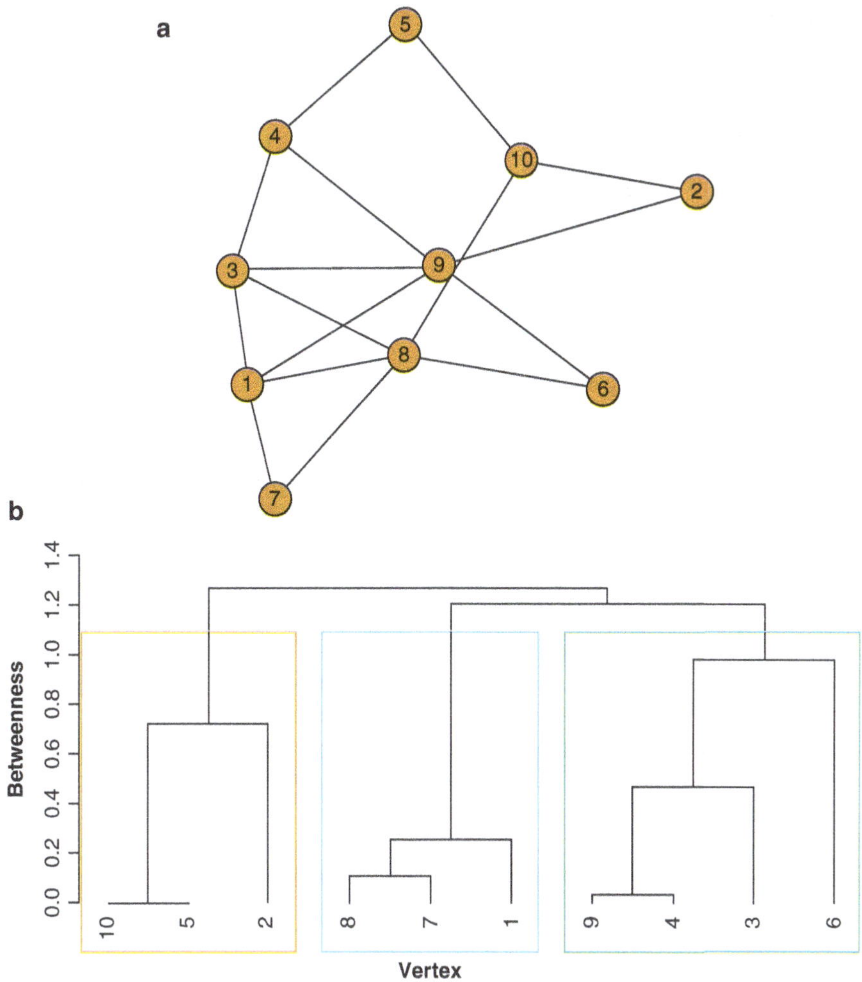

Fig. 13.2 Example of a graph (**a**) and a corresponding hierarchical clustering dendogram (**b**)

vertices are together. As it works down the tree, branches split off and community structures are formed as seen by the squares that group and encircle different vertices. Figure 13.2b shows a cutoff betweenness of approximately 1.1. This is seen by the clustering rectangles reaching a betweenness of 1.1 on the y axis.

It is important to note that betweenness centrality clustering can be performed on a graph with edge weights. By default, a graph assumes a topological edge weight of 1 as described in a previous section of this chapter. The way the GN algorithm functions is that in calculating edge betweenness the algorithm multiplies the betweenness of an edge by the weight of that edge in determining edges with the highest overall betweenness considering the weight of the edges. In this work, topological, admittance, impedance, and length weights were used in conjunction with betweenness centrality graph clustering.

2. **Label Propagation Graph Clustering (LPGC):** Another useful algorithm for determining community structure in networks that was used in this work was an algorithm called "Label Propagation" [New04]. The core idea for label propagation graph clustering (LPGC) is for a set of node labels to "propagate" through a network such that each node is assigned a label that corresponds to the nodes of its neighbors. One of the main advantages of this algorithm is that its computation time is near linear. For large systems, this algorithm provides fast computation. The general process for pseudocode of the LPGC is as follows [New04]:

   (a) Assign a unique label to each node.
   (b) Reassign node labels, node will be assigned the label that most of its neighbors are labeled with.
   (c) If ties occur, they are broken randomly
   (d) Stoppage criterion occurs when every node in the network has a label to which the maximum number of its neighbors belong to. This stoppage criterion requires that each vertex has at least as many neighbors within its community as it has with each of the other communities.

3. **Nearest-Generator (NG) Clustering:** Nearest-generator clustering is a simple but novel method utilized in this work specifically due to the requirements of power systems. The nearest-generator method is appropriately named as the algorithm functions by assigning each bus in the system to the generator which it is nearest to according to a desired edge-weight metric such as line length or impedance. This method was developed for a few reasons. The first reason is the trivial logic of assigning a demand bus to the generator that it is nearest to. The second reason this method was developed was that it is an efficient way to ensure that the cluster decompositions follow the microgrid rule of containing at least one generator. The authors know of no graph clustering algorithm that by default would cluster the bus system in a way where each cluster would contain at least one generator.

4. **Two-Stage Graph Clustering Method:** To adjust for scalability of bus systems as they get larger, a two-stage method of graph clustering was applied. As the size of the bus system increases, the number of clusters formed by a graph clustering algorithm will also increase as a general rule. As an example, when betweenness centrality clustering is applied to the IEEE 300-bus system 14 clusters result as the scheme with optimal modularity using this algorithm. In order to decrease the number of clusters while still respecting optimal modularity and improve the certainty that each cluster will contain generation and load, a two-stage clustering method was adopted.

   The general process of a two-stage method is as follows: A graph clustering algorithm is applied to a desired network graph with the stoppage criterion of the algorithm being set to optimal modularity. Once the algorithm has computed a community structure, the topology of this structure is converged. A converged community structure is essentially treating the output memberships of a graph clustering algorithm as new graph vertices. As a simple example, Fig. 13.3 contains nine vertices. A graph clustering algorithm is applied resulting in three microgrids as shown by the three clusters in the first-stage method.

   A converged graph of this community structure assumes each community of vertices output from a graph clustering algorithm to be a single vertex in a new representative graph. Additionally, the edges between communities are the only

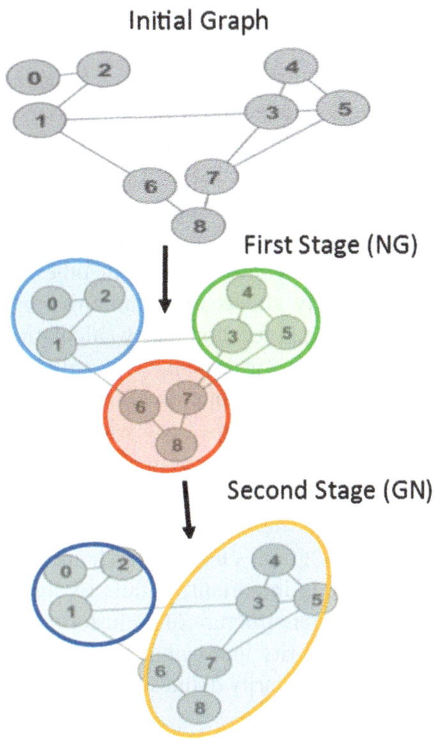

Fig. 13.3 Two-stage process

edges considered in a converged graph. Figure 13.3 contains a flowchart that describes the two-stage process. The application of the first clustering algorithm results in a three-microgrid system by clustering the nine-bus system. From this converged graph, an additional graph clustering algorithm is applied, thus becoming a two-stage method. A second-stage algorithm is applied to the converged clusters from the first-stage method resulting in a final cluster formation containing only two clusters that represent and fully contain the original nine vertices.

Deploying a two-stage method allows for important desired results to be achieved. One consequence of a two-stage method is that the overall number of clusters can be reduced when the system is large. Another important characteristic of two-stage clustering is that desirable attributes of multiple clustering algorithms can be considered in a single clustering scheme. As an example, in this work, a combination of nearest-generator clustering and betweenness centrality clustering were used in a two-stage method.

Two-Stage NG-GN: The nearest-generator (NG) method was used as a first-stage algorithm and Girvan–Newman (GN) as the second stage. Using these two algorithms in a two-stage method ensures that positive attributes of both algorithms are encapsulated in the final result. The use of NG as the first-stage method ensures that every cluster will have a generator. Using weighted betweenness as a second-stage algorithm will encorporate the utility of finding community structure based on betweenness. This method was deployed on the three-bus systems and the economic dispatch for each was analyzed.

5. **Graph Cluster Modularity**: A method for quantifying the strength of a graph clustering decomposition is necessary in order to quantitatively understand how well a graph decomposition is actually clustered. This is why there is need for graph modularity [SFO13], [GI16]. Modularity index measures the strength of a division of a graph into clusters. Cluster decompositions with high modularity scores have dense connections between vertices within clusters but sparse connections between vertices in other clusters. The calculation of graph cluster modularity allows for quantitative optimization of a graph clustering scheme. The formula for modularity calculation is as follows.

   A clustering scheme will split a graph into a division with k number of clusters. The calculation of modularity first involves the construction of a matrix $\boldsymbol{e}$ with dimensions $k \times k$ whose element e_{ij} is the fraction of all edges in the graph that link vertices in cluster i to vertices in cluster j. Conversely, the trace of this matrix, $Trace(\boldsymbol{e}) = \Sigma_i e_{ii}$, is the fraction of edges in the graph that connect vertices in the same cluster. The trace has a maximum of $Trace(\boldsymbol{e}) = 1$. In an efficient graph clustering scheme, the trace is ideally near to 1. While this number is important, it fails to signify any information about connections to intercluster structure of a clustering scheme [GN03], [New06], [SFO13].

   Modularity index goes another step further in including intercluster connections. Modularity defines a row sum $a_i = \Sigma_j e_{ij}$ that represent the fraction of edges that connect to vertices in cluster i. Regarding these values, modularity is calculated by:

$$Q = \Sigma_i \left(e_{ii} - a_i^2\right) = Trace(e) - \parallel e^2 \parallel \tag{13.1}$$

where $\|x\|$ indicates the sum of the elements of the matrix x. This value measures the fraction of edges in the graph that connect vertices of the same cluster minus the expected value of the same quantity in a network with the same community divisions but random connections between the vertices. This metric essentially compares the connections of one scheme to the same scheme with the same number of random connections. If the number of within-cluster edges is no better than random, then $Q=0$. The maximum value of Q is 1. Numbers near 1 indicate stronger cluster structure. In practice, values for networks will typically fall between 0.3 and 0.7. Higher values are considered to be rare [SFO13], [GI16].

13.2.3 Software Utilization

The authors used a statistical software, R, and a mathematical programing software, AMPL to implement the graph-based clustering techniques and economic dispatch linear program. R is an open source software programming environment that contains an integrated suite for calculation, graphical display, and data manipulation [UWD16]. AMPL stands for "A Mathematical Programming Language" and is a useful software for performing optimization using linear programming.

13.3 Implementation and Modeling

13.3.1 Procedure

Figure 13.4 shows the different steps followed for the cluster analysis. A suitable graph clustering scheme will be applied to a standard bus system to obtain the different clusters. A multi-area economic dispatch will be applied to the clusters to do the cost analysis and different indices will also be used to compare the clusters.

In this chapter, static edge weight methods were examined in conjunction with betweenness centrality and two-stage graph clustering as they were applied to the IEEE 57, IEEE 118, and IEEE 300 bus test systems. These methods were applied to intelligently form power network zones or microgrid type structures within the test bed system. IEEE bus test system data was obtained from [HSA12]. Test system data come in the form of two separate spread sheets. One sheet contains data pertaining to information regarding the buses (vertices) in the system. The other

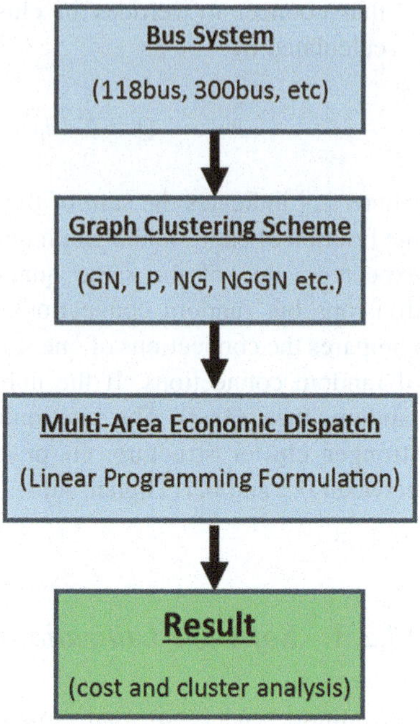

Fig. 13.4 The steps followed for the cluster analysis

spreadsheet contains data regarding the transmission lines (edges) in the system. Data sheets were analyzed using the "igraph" package of R software.

Transmission line lengths were calculated according to the method proposed by [Fre77]. Admittance/impedance weights of each transmission line were calculated per the specified line resistance and reactance. BC was also calculated for each system. After calculation of the weights, graph clustering per weighted betweenness centrality was performed using the GN algorithm for each type of weight (topological, admittance/impedance, and length) for all three-bus systems. In addition to BCGC, other graph clustering algorithms were deployed for grid decomposition. Alternate methods included combinations of a two-stage clustering algorithm and the nearest-generator method.

The nearest-generator method was deployed both by itself and as a first stage of the two-stage clustering method. The nearest-generator method assigns each load to the generator that it was nearest to in accordance with the edge weights (e.g., impedance of a transmission line) with the stipulation that each bus on its path to the nearest generator was also assigned to the same microgrid. This stipulation was added because although a bus "*a*" may be nearest to generator "*x*", but this does not guarantee that the buses along bus *a*'s path to *x* are also most near to that generator. Ties and differences may occur. The other deployed graph clustering algorithms do not guarantee that the cluster scheme forms viable microgrids as it is possible, for example, that some of the clusters/microgrids formed do not contain any generators. This was the main reason for why the nearest-generator method was deployed. In this case the decomposition is considered infeasible because it does not follow the definition of a microgrid in its decomposition. We denote this by identifying if the decomposition follows microgrid rules (MGRs). To quantify the effectiveness of the microgrid decomposition, graph cluster modularity was calculated and compared for each decomposition. The bus system graphs were then plotted as network flow graphs using software capabilities to achieve an effective visualization.

Additionally, to adjust for scalability of bus systems as they get larger the two-stage method of graph clustering was applied. In this work the two-stage combination of the nearest-generator graph clustering algorithm and the Girvan–Newman edge betweenness algorithm were deployed. The two-stage clustering method could theoretically involve any combination of algorithms. This work presents a two-stage method of nearest generator (NG) and Girvan–Newman (GN). The NG graph clustering algorithm was deployed as the first algorithm of the two-stage process. The GN algorithm was deployed as the second-stage algorithm. Together the combination of these two algorithms in the two-stage clustering process was denoted NGGN. Other combinations of algorithms were used in the two-stage process, but since none of them by default guarantee that MGRs will be followed, the NGGN method was analyzed most thoroughly.

13.4 Numerical Simulations

To see the impact of decomposition structures on cost, an economic load dispatch problem is applied to these IEEE test systems. Economic load dispatch (ELD) is a method to schedule the power generator outputs with respect to the load demands and to operate the power system most economically. In other words, the main objective is to allocate the optimal power generation of different units at the lowest cost possible while meeting all system constraints [SBA13]. The economic load dispatch is performed in a multigenerator system to schedule the generators to satisfy the loads in the system subjected to generator and transmission line limits. Attaining an optimal point in the generation values will result in significant savings to power system companies in terms of fuel cost and resource utilization. In power system, minimization of operation cost is very important, therefore we can use ELD as an effectiveness way to evaluate the different clustering techniques.

The clustering techniques divide the bus system into different zones or areas and the application of economic dispatch on such a system is known as Multi-Area Economic Dispatch (MAED). The aim of MAED problems is the minimization of power generation cost while satisfying the load demand in the system and subject to generation and line flow constraints. The fuel cost for a generating unit i (*in \$ per hour*) supplying P_{Gi} amount of real power can be represented by a quadratic equation [HSA12] as shown in Eq. (13.2):

$$F_i(P_{\mathrm{Gi}}) = a_i {P_{\mathrm{Gi}}}^2 + b_i P_{\mathrm{Gi}} + c_i \tag{13.2}$$

where a_i, b_i, and c_i are the cost-coefficients of generating unit i and P_{Gi} is the real power generation of the unit i. The objective is to minimize the total cost of generation, which can be represented by the following equation:

$$F = \sum_{i=1}^{n_g} F_i(P_{\mathrm{Gi}}) \tag{13.3}$$

where n_{g} is the number of generators working in the bus system. The economic dispatch problem is then solved subject to several formulated constraints [ZRK07]. They are listed in Eqs. (13.4)–13.7).

$$P_{\mathrm{Gi}_{\min}} \leq P_{\mathrm{Gi}} \leq P_{\mathrm{Gi}_{\max}} \quad \text{for} \quad i = 1....n_{\mathrm{g}} \tag{13.4}$$

$$\sum_{i=1}^{n_g} P_{\mathrm{Gi}} = D \tag{13.5}$$

Here, the constraint (13.4) implies that the generation from each generator must be within its maximum and minimum values and the constraint (13.5) shows the condition that the generation from all the generators should meet the total demand in the system. The power flow through the tie lines connecting the areas is an additional constraint in the MAED problem, as shown in constraint (13.6).

$$T_{mn_{\min}} \leq T_{mn} \leq T_{mn_{\max}} \tag{13.6}$$

The power flow between two areas m and n is subjected to a minimum and a maximum value of $T_{mn\min}$ and $T_{mn\max}$ respectively.

$$\sum_{i=1}^{m_g} P_{\mathrm{Gi}} - \sum_{j=1}^{t_c} T_{\mathrm{cj}} + \sum_{k=1}^{t_c} T_{\mathrm{kc}} = D_c \tag{13.7}$$

Equation (13.7) ensure that the loads in each of the zones are satisfied by the generation within the zone and from the power flow from neighboring zones. Here, m_{g} indicates the number of generators within the zone, t_{c} the number of tie lines connected to the zone and T_{cj} and T_{kc} indicate the power flowing out from the zone and coming to the zone from connected zones, since the analysis is done considering bi-directional power flow between the clusters. The variable D_{c} indicates the total active load in the microgrid under consideration. In our model, the cost of power flow through the tie lines is also taken into consideration. A value of 0.1$ per MW is taken as its cost and a 200 MW power limit is applied as maximum tie line flow limit [SCJ13]. In order to compare the different zones obtained by using the different clustering techniques, the generator cost functions are assumed to be the same for all the generators in the grid system. The total cost function to be minimized will be the sum of generation cost plus the cost of tie line power flow and the modified equation is given in (13.8).

$$F = \sum_{i=1}^{n_g} F_i(P_{Gi}) + \sum_{j=1}^{t} C_j T_j \tag{13.8}$$

The variable C_j denotes the cost of tie line power flow which is assumed to be constant for all the tie lines, t represents the number of tie lines in the model, and T_j is the amount of tie line power flow [Dan95]. The ED model is developed using the concept of linear programming with the above-mentioned load, generator and tie line flow constraints. The ED model is programmed using AMPL (A Mathematical Programming Language), it is a popular tool used for solving linear programming problems. AMPL software needs two type of files: a model and data file. The mod file contains the linear programming code and .dat file contains the information or data the code works on. A separate .dat and .mod files are created for IEEE 57-, 118- and IEEE 300-bus systems for every decomposition structures. The ED model considers only the active loads and generators in the system and does not consider the reactive power in the system. Tables 13.1 and 13.2 list the generation cost, tie line flow cost and the total cost for the IEEE 118- and 300-bus systems for the Length-GN (L-GN), Admittance-GN(A-GN) and NGGN clustering technique respectively.

For the 118-bus system, there is a 66.6% in reduction in tie-line flow cost for the A-GN clustered system when compared to the L-GN system and there is a significant reduction of 86.13% in the case of NGGN method when compared to the L-GN method. In the case of total cost the cost reductions are 1.21% and 1.64% respectively for the A-GN and NGGN method.

For the 300-bus system case, the values for the tie line flow cost reductions are 0% for the A-GN method since the cluster was identical to the L-GN method and it is 29.91% for the LPGN. Similarly, the reduction in total costs are 0% for the A-GN and 12.64% for the NGGN method. From these results, there is a significant reduction in tie line flow cost for the LPGN clustering technique when compared

Table 13.1 ED cost distribution in an IEEE 118-bus system

	L-GN	A-GN	NGGN
Number of clusters	10	11	5
Generation cost ($)	9074.86	9137.03	9148.06
Tie line flow cost ($)	262.871	87.725	36.45
Total Cost ($)	9337.73	9224.76	9184.51

Table 13.2 ED cost distribution in an IEEE 300-bus system

	L-GN	A-GN	NGGN
Generation cost ($)	141,704	141,704	123,824
Number of clusters	14	14	8
Tie line flow cost ($)	233.089	233.089	163.361
Total cost ($)	141,937	141,937	123,988

Table 13.3 Generator/load ratio of the 118-bus system clusters

Generation/load ratio	L-GN	A-GN	NGGN
Maximum value (%)	198.61	277.08	124.48
Minimum value (%)	49.62	47.22	50.63

Table 13.4 Generator/load ratio of the 300-bus system clusters

Generation/load ratio	L-GN	A-GN	NGGN
Maximum value (%)	616.61	616.61	411.86
Minimum value (%)	61.90	61.90	86.96

to L-GN and A-GN techniques and using the NGGN method we are also able to achieve a reduction in total cost of the system.

Another parameter which can be used to compare grid clusters is the generation to load (G/L) ratio in each of the individual clusters. Table 13.3 list the maximum and minimum value of this G/L for the IEEE 118-bus system, excluding the zones with no active power generation. A G/L value that is more than 100% indicates a self-sufficient grid cluster with excess generation that can be given to other microgrids. A G/L value less than 100% indicates that the generation within the cluster is not sufficient to satisfy its load and thus require resources from neighboring microgrids to meet its demand. Table 13.4 lists the G/L ratio for the IEEE 300-bus systems and from both these results LPGN clusters are more suited for a grid system when compared to either of the other two cases because the values are closer to the ideal value of 100.

13.5 Results

Several different decomposition criteria were utilized in analyzing the different grid decompositions. This section shows a sample of some effective techniques. An exhaustive display of structures is not provided due to length of the chapter. Modularity scores for these criteria are recorded and shown in Tables 13.5, 13.6 and 13.7.

These tables show a modularity score, and whether the given decomposition follows rules for being considered a microgrid. Modularity index for microgrid decomposition is a useful metric in determining a grid structures ability to withstand microgrid or cascading failures. High modularity indicates dense microgrid intra-connection while simultaneously maintaining sparse interconnection with other microgrids. The physical bus system decomposition structures for the 118- and 300-bus systems that accompany these tables can be seen by the visualizations contained in Fig. 13.5. Tables 13.8 and 13.9 list all buses in their respective zones obtained using the three clustering techniques for the IEEE 118- and 300-bus systems.

Table 13.5 IEEE 57-bus system displaying modularity scores of microgrid decomposition as well as displays whether the decomposition follows valid microgrid rules and is a feasible decomposition

IEEE 57-bus test system		
Cluster scheme	Modularity	Follows MGR
Topology/GN	0.6307	No
L-GN	0.5505	No
A-GN	0.5071	No
Topology-NG	0.427	Yes
Length-NG	0.4424	Yes
Admittance-NG	0.45	Yes
NGGN	0.434	No

Table 13.6 IEEE 118-bus system with modularity scores and the algorithm that obeys microgrid rule (MGR) for microgrid decomposition

IEEE 118-bus test system		
Cluster scheme	Modularity	Follows MGR
Topology-GN	0.6908	Yes
L-GN	0.6721	No
A-GN	0.74537	Yes
Topology-NG	0.5151	Yes
Length-NG	0.2995	Yes
Admittance-NG	0.1312	Yes
NGGN	0.6644	Yes

Table 13.7 IEEE 300-bus system displaying modularity scores of microgrid decomposition as well as displays whether the decomposition follows valid microgrid rules and is a feasible decomposition

IEEE 300-bus test system		
Cluster scheme	Modularity	Follows MGR
Topology-GN	0.8344	No
L-GN	0.7824	No
G-GN	0.8344	No
NGGN	0.784	Yes

13.6 Discussion

Our preliminary results indicate that the highest modularity score for either test bus system occurred using the GN algorithm with edge weights weighted with admittance of the transmission lines. This was evident in the 118-bus system. The 57-bus system microgrid decompositions were infeasible, when clustered according to any weighted GN algorithm. The bare topology of the 57-bus system does not respect betweenness clustering because generators in the 57-bus system are not well distributed and are topologically positioned very near to each other. This causes betweenness graph clustering to, by default, form microgrids that do not have generators within them. Thereby, the decompositions contain microgrids that do not hold to the definition of microgrid and are hence infeasible. However, when using the nearest-generator algorithm in the 57-bus system, the highest modularity

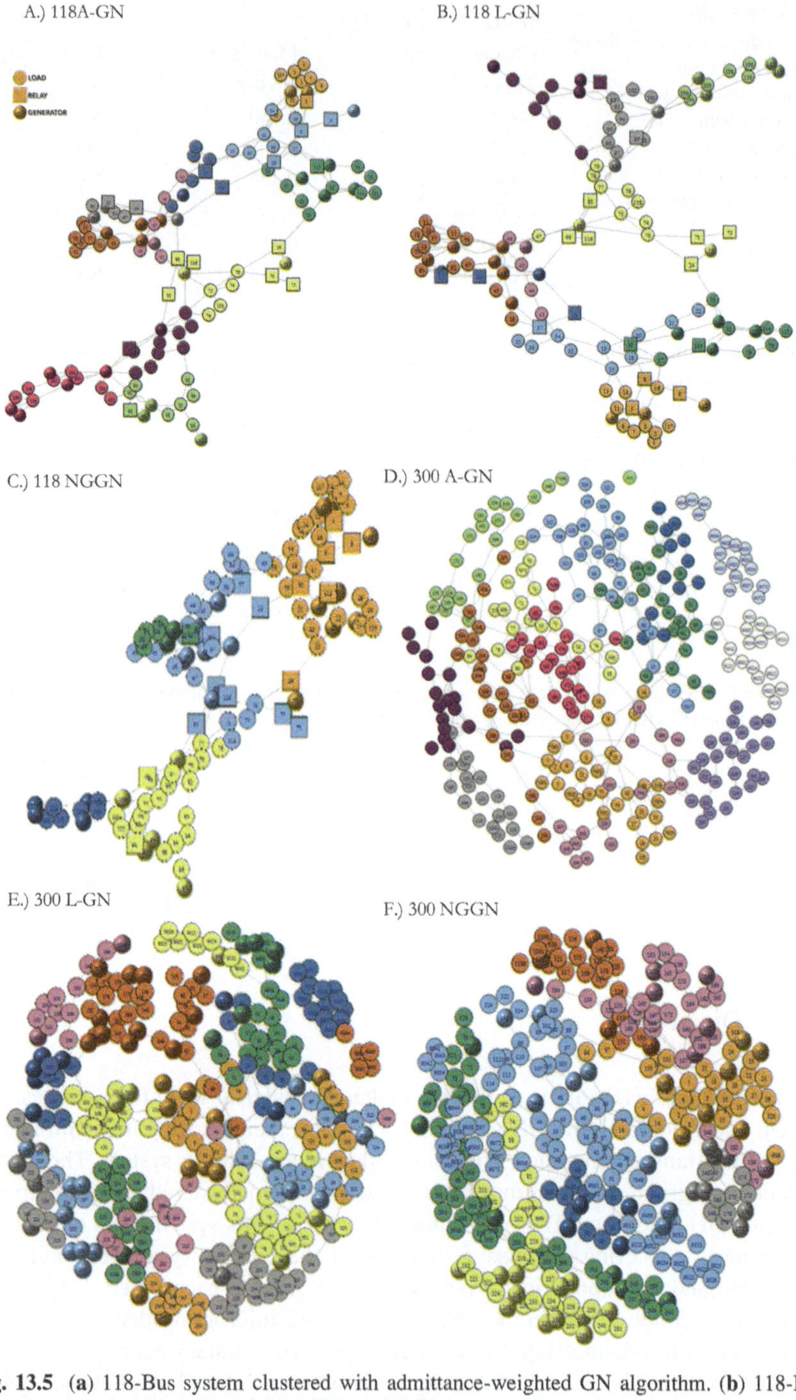

Fig. 13.5 (**a**) 118-Bus system clustered with admittance-weighted GN algorithm. (**b**) 118-Bus system clustered with length-weighted GN. (**c**) 118-Bus system clustered with two-stage NG + GN. (**d**) 300-Bus system clustered with admittance-weighted GN. (**e**) 300-Bus system clustered with length-weighted GN. (**f**) 300-Bus system clustered two-stage NGGN

Table 13.8 IEEE 118-bus grid decomposition

	L-GN	A-GN	LPGN
Zone 1	1,2,3,4,5,6,7,8,9,10,11,12, 13,14,16,17,117	1,2,3,4,5,6,11,12,13, 117	1,2,3,4,5,6,7,8,9,10,11,12,13,14,16,117
Zone 2	8,9,10,14,16,17,18,19,30,33	17,23,25,26,27,28,29,30,31, 32,113,114,115	15,17,18,19,20,21,22,23,24,25,26,27,28, 29,30,31,32,33,113,114,115
Zone 3	15,18,19,20,21,22,33,34,35,36,37	20,21,22,23,25,26,27,28,29,31, 32,113,114,115	34,35,36,37,38,39,40,41,42,43,44,45,46,47,48,49, 50,51,52,53,54,55,56,57,58
Zone 4	24,47,68,69,70,71,72,73,74, 75,76,77,78,79 ,81,116,118	24,68,69,70,71,72,73,74,75,76,81,116,118	59,60,61,62,63,64,65,66,67,68,69,70,71,72,73, 74,75,76,81,116,118
Zone 5	82,83,84,85,86,87,88,89,90,91	34,35,36,37,38,39,40,41	77,78,79,80,82,83,84,85,86,87,88,89,90,91,92, 93,94,95,96,97,98,99,100,101,102,103
Zone 6	38,61,63,64,65	43,44,45,46,47,48	104,105,106,107,108,109,110,111,112
Zone 7	43,44,45,46,48	42,49,50,51,52,53,54,55,56,57,58,66	NA
Zone 8	39,40,41,42,49,50,51,52,53, 54,55,56,57,58,59,60, 62,66,67	59,60,61,62,63,64,65,67	
Zone 9	80,92,93,94,95,96,97,98,99, 100,101,102	77,78,79,80,82,94,95,96,97,98,99	
Zone 10	103,104,105,106,107,108,109, 110,111,112	100,101,102,103,104,105,106,107, 108,109,110,111,112	
Zone 11	NA	83,84,85,86,87,88,89,90,91,92,93	

Table 13.9 IEEE 300-bus grid decomposition

	L-GN	A-GN	NGGN
Zone 1	1,2,3,4,5,6,7,8,9,10,11,12,13, 14,16,19,20,21,22,23,24,25,26,27, 319,320,7001,7002,7003,7011, 7012,7023	1,2,3,4,5,6,7,8,9,10, 11,12,13,14,16,19,20,21,22,23,24,25,26, 27,319,320,7001,7002,7003, 7011,7012,7023	1,2,3,4,5,6,7,8,9,10,11, 12,13,14,15,16,17,19,20,21,22, 23,24,25,26,27,128,129,130,131, 150,167,168,319,320,7001,7002, 7003,7011,7012,7017,7023,7024,7130
Zone 2	15,17,47,85,86,87,89,90,91, 92,94,97,98,99,100,102,103,104,105, 107,108,109,110,112,113, 114,322,323,324,7017	15,17,47,85,86,87,89,90,91,92,94, 97,98,99,100,102,103,104,105,107, 108,109,110,112,113, 114,322,323,324,7017	33,34,35,36,37,38,39,40,41,42, 43,44,45,46,47,48,49,51,52,53, 54,55,69,70,71,72,73,74,76,77,78, 79,80,81,84,85,86,87,88,89,90,91, 92,94,97,98,99,100,102,103,104,105,107, 108,109,110,112,113,114,189,193,195, 196,197,198,199,200,201,202,203,204, 205,206,207,208,209,210,211,212, 322,323,324,528,531,552,562,609, 2040,7039,7044,7049,7055,7071
Zone 3	33,34,37,38,39,40,41,42,43,44,45,46, 48,49,51,52,53,54,55,7039,7044, 7049,7055	33,34,37,38,39,40,41,42,43,44,45, 46,48,49,51,52,53,54,55,7039,7044, 7049,7055	57,58,59,60,61,62,63,64,526, 7057,7061,7062
Zone 4	35,36,70,71,72,73,74,76,77,78, 80,84,88,528,531,552,562,609,7071	35,36,70,71,72,73,74,76,77,78,80,84,88, 528,531,552,562,609,7071	115,116,117,118,119,120,121,122,123, 124,125,126,127,132,133,134,135,136,137 ,138,151,152,153,154,155,156,157,158,159, 160,161,162,163,164,165,166,169,170,171,181, 183,184,185,186,187,188,1190,1200,1201,7166
Zone 5	57,58,59,60,61,62,63,64,526,7057, 7061,7062	57,58,59,60,61,62,63,64,526,7057, 7061,7062	139,140,141,142,143,144,145,146,147, 148,149,172,173,174,175,176,177,178,179, 180,182,7139
Zone 6	69,79,189,193,196,197,198,199,200, 201,202,203,204,205,206,207,208,209, 210,211,248,249,250,2040	69,79,189,193,196,197,198,199,200,201, 202,203,204,205,206,207,208,209,210, 211,248,249,250,2040	190,191,192,194,216,217,218,219,220, 221,222,223,224,225,226,227,228,229,230,231, 232,233,234,235,236,237,238,239, 240,241,281,664

Zone 7	81,194,195,212,213,214,215,216, 217,218,219,242,243,244,245,246, 247,664	81,194,195,212,213,214,215,216, 217,218,219,242,243,244,245,246, 247,664	213,214,215,242,243,244, 245,246,247,248,249,250
Zone 8	115,116,117,118,119,120,121,122, 123,124,125,126,157,158,159,160, 1190,1200,1201	115,116,117,118,119,120,121,122,123, 124,125,126,157,158,159,160,1190, 1200,1201	9001,9002,9003,9004,9005,9006,9007,9012, 9021,9022,9023,9024,9025,9026,9031,9032, 9033,9034,9035,9036,9037,9038,9041,9042, 9043,9044,9051,9052,9053,9054,9055,9071, 9072,9121,9533
Zone 9	127,128,129,130,131,132,133,134, 135,150,151,167,168,169,170, 171,184,185,7130	127,128,129,130,131,132,133,134,135, 150,151,167,168,169,170,171,184, 185,7130	NA
Zone 10	136,137,138,152,153,154,155,156, 161,162,163,164,165,166,181,183,186, 187,188,7166	136,137,138,152,153,154,155,156,161, 162,163,164,165,166,181,183,186,187, 188,7166	
Zone 11	139,140,141,142,143,144,145,146,147, 148,149,172,173,174,175,176,177,178, 179,180,182,7139	139,140,141,142,143,144,145,146,147, 148,149,172,173,174,175,176,177,178, 179,180,182,7139	
Zone 12	190,191,192,220,221,222,223,224, 225,226,227,228,229,230,231,232,233, 234,235,236,237,238,239,240,241,281	190,191,192,220,221,222,223,224,225, 226,227,228,229,230,231,232,233,234,235, 236,237,238,239,240,241,281	
Zone 13	9001,9002,9005,9012,9021,9022, 9023,9024,9025,9026,9051,9052,9053, 9054,9055,9121,9533	9001,9002,9005,9012,9021,9022,9023, 9024,9025,9026,9051,9052,9053, 9054,9055,9121,9533	
Zone 14	9003,9004,9006,9007,9031,9032,9033, 9034,9035,9036,9037,9038,9041,9042, 9043,9044,9071,9072	9003,9004,9006,9007,9031,9032,9033, 9034,9035,9036,9037,9038,9041,9042, 9043,9044,9071,9072	

Note: The numbering of IEEE 300-bus do not follow an ordered sequence after bus number 17

score occurs when the edge weights are determined using admittance. The nearest-generator (NG) algorithm works poorly when applied by itself in the 118- and 300-bus systems. Betweenness centrality is a good metric for determining microgrid structures within a given grid system, if generators are well distributed. This is likely to be the case for smart grid networks. Further work needs to be done to observe the scalability of these algorithms and focus could be attended to development of a more specific algorithm that considers microgrid rules and the BC of objects in the system. An algorithm considering these factors would provide decompositions with high modularity but also conform to definitions of microgrid. Modularity index for microgrid decomposition is a useful metric in determining a grid structure's ability to withstand microgrid or cascading failures. High modularity indicates dense microgrid intra-connection while simultaneously maintaining sparse interconnection with other microgrids. However, the modularity score of a decomposition does not appear to have a significant relationship with the optimization of economic dispatch.

13.7 Conclusion

When examining economic dispatch of different microgrid decompositions, analysis shows that decompositions formed using two-stage clustering method show a reduction in cost. This reduction in cost is mostly due to savings that occur in tie line flow. Savings that occur in generation cost are quite small and not statistically significant. The reduction in tie line flow cost results from more even distribution of tie line flows found in the decompositions formed by two-stage clustering. Though modularity for two-stage clustering schemes was slightly lower than one-stage schemes, the economic dispatch showed to be more cost-effective. Though modularity is a good indicator of clustering efficiency, its effect on economic dispatch is less clear. The two-stage clustering method making use of admittance/impedance weighted betweenness in conjunction with the nearest-generator method is a novel method of analysis and grid decomposition. This method shows an overall reduction in dispatch cost compared to single-stage clustering methods while simultaneously ensuring that each microgrid zone will contain generation. Further work involving scaling this method to very large systems and comparative analysis with existing power system bus decompositions will add value.

References

[AAT05] Alekhnovich, M., Arora, S., & Tourlakis, I. (2005). Towards strong non approximability results in the Lovasz-Schrijver hierarchy. *Theory of computing* (pp. 294–303). ACM.

[ABL+06] Arora, S., Bollobas, B., Lorasz, L., & Tourlakis, I. (2006). Proving integrality gaps without knowing the linear program. *Theory of computing* (pp. 19–51).

[ADH+94] Adler, R., Daniel, S., Heising, C., Lauby, M., Ludorf, R., & White, T. (1994). An IEEE survey of US and Canadian overhead transmission outages at 230 kV and above. *IEEE Transaction on Power Delivery, 9*(1), 21–39.

[Ami04] Amin, M. (2004). Balancing market priorities with security issues: Interconnected system operations and control under the restructured electricity enterprise. *IEEE Power and Energy Magazine, 2*(4), 30–38.

[Ami05] Amin, M. (2005). Powering the 21st century: We can and must modernize the grid. *IEEE Power and Energy Magazine*, 93–95.

[Pha93] Phadke, A. G. (1993). Synchronized phasor measurements in power systems. *IEEE Computer Applications in Power, 6*(2), 10–15.

[AMP] *Advanced modeling and programming language.* [Online]. Retrieved November 5, 2014, from http://www.ampl.com/.

[Aro02] Arora, S. (2006). Proving integrality gaps without knowing the linear program. *Proceedings of the 43rd Symposium on Foundations of Computer Science* (pp. 313–322).

[AS08] Amin, M., & Schewe, P. (2008). Preventing blackouts. *Scientific American, 296*, 60–67.

[AW05] Amin, M., & Wollenberg, B. (2005). Toward a smart grid. *IEEE Power and Energy Magazine, 3*(5), 34–38.

[Arm85] Armstrong, J. S. (1985). *Long-range forecasting: From crystal ball to computer.* New York, NY: Wiley.

[Abu04] Abur. (2004). Observability analysis and measurement placement for systems with PMUs, *IEEE PES Power Systems Conference and Exposition*, (1), 1472–1475.

[AC92] Armstrong, J. S., & Collopy, F. (1992). Error measures for generalizing about forecasting methods: Empirical comparisons. *International Journal of Forecasting, 8*, 69–80.

[BB91] Balakrishnan, B., & Balemi, S. (1991). Branch and bound algorithm for computing the minimum stability degree of parameter-dependent linear systems. *International Journal of Robust and Nonlinear Control, 1*(4), 295–317.

P. Ranganathan, K.E. Nygard, *Distributed Linear Programming Models in a Smart Grid*, Power Electronics and Power Systems, DOI 10.1007/978-3-319-52617-1

[BBG77] Bradley, G., Brown, G., and Graves, G (1977). Design and implementation of large scale primal transshipment algorithms. *Management Science*, 1–34.

[BCP08] Bignucoloa, F., Caldona, R., & Prandonib, V. (2008). Radial MV networks voltage regulation with distribution management system coordinated controller. *Electric Power Systems Research, 78*(4), 634–664.

[BF96] Bruha, I., & Franek, F. (1996). Comparison of various routines for unknown attribute value processing: The covering paradigm. *International Journal of Pattern Recognition and Artificial Intelligence, 10*(8), 939–955.

[BHP09] Blumsack, S., Hines, P., Patel, M., Barrows, C., & Sanchez, E. C. (2009). Defining power network zones from measures of electrical distance. *2009 I.E. Power and Energy Society General Meeting, PES* (pp. 1–8).

[BJN98] Barnhart, C., Johnson, E., & Nemhauser, G. (1998). Branch-and-price: Column generation for solving huge integer programs. *Operations Research, 46*(3), 316–329.

[BJS10] Bazaraa, M. S., Jarvis, J. J., & Sherali, H. D. (2010). *Linear programming and network flows*. Hoboken, NJ: Wiley.

[Bou01] Boutilier, C. (2001). *Planning and programming with first-order Markov decision processes: Insights and challenges*. City, MA: Morgan Kaufmann.

[Bou02] Boutilier, C. (2002). A POMDP formulation of preference elicitation problems. *Proceedings of the Eighteenth National Conference on Artificial Intelligence*, City, CA (pp. 239–246).

[BMB+93] Baldwin, T. L., Mili, L., Boisen, M. B., & Adapa, R. (1993). Power system observability with minimal phasor measurement placement. *IEEE Transactions on Power Apparatus and Systems, 8*(2), 707–715.

[Mar94] Marick, B. (1994). *The craft of software testing: Subsystem testing including object-based and object-oriented testing*. Upper Saddle River, NJ: Prentice-Hall, Inc.

[BYA] Bei, X., Yoon, Y. J., & Abur, A. *Optimal placement and utilization of phasor measurements for state estimation*. Texas A&M University. Technical Report.

[BW03] Bailey, D., & Wright, E. (2003). *Practical SCADA for industry*. Oxford, England: Newnes.

[CB02] Boutilier, C. (2002). A POMDP formulation of preference elicitation problems. *Proceedings of the Eighteenth National Conference on Artificial Intelligence*.

[Chv83] Chvatal, V. (1983). *Linear programming*. New York, NY: Freeman.

[CI94] Cavalier, T. M., & Ignizio, J. P. (1994). *Linear programming*. Englewood Cliffs, NJ: Prentice-Hall, Inc.

[CKE09] Chakrabarti, S., Kyriakides, E., & Eliades, D. G. (2009). Placement of synchronized measurements for power system observability. *IEEE Transactions on Power Delivery, 24*(1), 12–19.

[CLD+02] Carreras, B., Lynch, V., Dobson, I., & Chaoes, N. (2002). Critical points and transitions in an electric power transmission model for cascading failure blackouts. *Chaos, 12*(4), 985–994.

[CT99] Camponogara, C., and Talukdar, S. (1999). Agent cooperation: Distributed control applications. *Proceedings of the International Conference on Intelligent System Application to Power Systems*, City, VA (pp. 1–6).

[Dij59] Dijkstra, E. (1959). A note on two problems in connexion with graphs. *Numerische Mathematik, 1*, 269–270.

[Dan51] Dantzig, G. (1951). Activity analysis of production and allocation. *Proceedings of a conference*, Cowles Commission for Research in Economics.

[Dan60] Dantzig, G. B. (1960). On the shortest path route through a network. *Management Science, 6*(2), 187–190.

[Dan63] Dantzig, G. B. (1963). *Linear programming and extensions*. Princeton, NJ: Princeton University Press.

[Dan83] Dantzig, G. B. (1983). *Reminiscences about the origins of linear programming, mathematical programming: The state of the art springer lecture notes* (pp. 78–86). Berlin, Germany: Springer.

[Dan95] Streiffert, D. (1995). Multi-area economic dispatch with tie line constraints. *IEEE Transactions on Power Systems, 10*(4), 1946–1951.

[Dan98] Dantzig, G. B. (1998). *Linear programming and extensions (# ed.)*. Princeton, NJ: Princeton University Press.

[DCN04] Dobson, I., Carreras, B., & Newman, D. E. (2004). Probabilistic load-dependent cascading failure with limited component interactions. *IEEE International Symposium on Circuits and Systems*, City, Canada (pp. 15–32).

[DD08] Dua, D., & Dambhare, S. (2008). Optimal multistage scheduling of PMU placement: An ILP approach. *IEEE Transactions on Power Delivery, 23*(4), 1812–1820.

[DDK+08] Dambhare, S., Dua, D., Kumar, R., & Soman, G. S. A. (2008). *Optimal zero injection considerations in PMU placement: An ILP approach*, 14–19.

[DES] U.S. Department of Energy. *Smart grid* [Online]. Retrieved from http://energy.gov/oe/technology-development/smart-grid.

[DLW+89] Dewitt, C., Lasdon, S., Waren, D., Brenner, A., & Melhem, A. (1989). OMEGA: An improved gasoline blending system for Texaco. *Interfaces, 19*(1), 85–101.

[DNS+95] Dwyer, A., Nielsen, R., Stangl, J., & Markushevich, N. (1995). Load to voltage dependency tests at B.C. hydro, *IEEE Transactions on Power Systems, 10*(2), 709–715.

[DOE] U.S. Department of Energy. *What is the smart grid*? [Online]. Retrieved from www.smartgrid.gov/the_smart_grid.

[FES12] Farag, E., El-Saadany, F., & Seethapathy, R. (2012). A two ways communication-based distributed control for voltage regulation in smart distribution feeders. *IEEE Transaction on Smart Grid*, 766–772.

[FF10] Ford, R., & Fulkerson, R. (2010). *Flows in networks*. Princeton, NJ: Princeton University Press.

[FG96] Franklin, S., & Graesser, A. (1996). Is it an agent, or just a program?: A taxonomy for autonomous agents. In *Proceedings of 3rd International Workshop on Agent Theories, Architectures and Languages* (pp. 21–35). NY: Springer.

[FGK03] Fourer, R., Gay, D. M., & Kernighan, B. W. (2003). *AMPL: A modeling language for mathematical programming* [Online]. Retrieved from http://www.ampl.com/.

[FGR09] Greitzer, F. L., Podmore R., Robinson M., & Ey, P. (2009). Naturalistic decision making for power system operators, *International Conference on Naturalistic Decision Making (NDM)*.

[Fil92] Fildes, R. (1992). The evaluation of extrapolative forecasting methods. *International Journal of Forecasting, 8*, 81–98.

[Fou83] Fourer, R. (1983). Modeling languages versus matrix generators for linear programming. *ACM Transactions on Mathematical Software, 9*(2), 143–183.

[FR12] Frank, S., & Rebennack, S. (2012). *A primer on optimal power flow: Theory, formulation, and practical examples*.

[Fre77] Freeman, L. C. (1977). A set of measures of centrality based on betweenness. *American Sociological Association, 40*(1), 35–41.

[Gie11] Giebel, G. (2011). *The state-of-the-art in short-term prediction of wind power – A literature overview* (2nd ed.).

[GAM00] Gou, B., Abur, A., & Member, S. (2000). A direct numerical method for observability. *IEEE Transactions on Power Systems, 15*(2), 625–630.

[GCR+13] Gopakumar, P., Chandra, G. S., Reddy, M. J. B., & Mohanta, D. K. (2013). Optimal redundant placement of PMUs in Indian power grid — northern, eastern and north-eastern regions. *Frontiers in Energy, 7*(3), 413–428.

[GH01] Grzymala-Busse, J. W., & Hu, M. (2001). A comparison of several approaches to missing attribute values in data mining. *LNAI, 2005*, 378–385.

[GI16] Gentleman, R. & Ihaka, R. (2016). *R: The R project for statistical computing*. R-project.org [Online]. Retrieved from https://www.r-project.org/.

[GK03] Gilmore, S., & Kloul, L. (2003). A unified approach to performance modeling and verification. *Paper presented at Dagstuhl seminar on Probabilistic Methods in Verification and Planning*. Germany.

[GN02] Girvan, M., & Newman, M. E. J. (2002). Community structure in social and biological networks. *Proceedings of the National Academy of Sciences of the United States of America, 99*(12), 7821–7826.

[GN03] Girvan, M. & Newman, M. E. J. (2003). *Finding and evaluating community structure in networks*. Cond-Mat/0308217, 1–16.

[GOB16] Guckenheimer, J., Overbye, T. J., Bienstock, D., Bose, A., Boston, T., Dagle, J., Ilic, M. D., Jones, C. K., F. P. Kelly, Y. G. Kevrekidis, R. D. Masiello, J. C. Meza, C. Rudin, R. J. Thomas, M. H. Wright, and Committee on Analytical Research Foundations for the Next-Generation Electric Grid. (2016). *Analytic research foundations for the next-generation electric grid*

[GPR+09] Greitzer, F., Podmor, R., Robinson, M., & Ey, P. (2009). Naturalistic decision making for power system operators. In *International Conference on Naturalistic Decision Making (NDM)*. London, England.

[GPS00] Gueret, C., Prins, C., & Sevaux, M. (2000). *Programmation lineaire*. Paris: Editions Eyrolles.

[GTC07] Grigsby, L. L., Tobergte, D. R., & Curtis, S. (2007). *Power systems* (Vol. 53(9), 2nd ed.). New York City: CRC Press.

[HL01] Hillier, F., & Lieberman, G. (2001). *Introduction to operations research*. City, England: McGraw-Hill.

[HRA+07] Hajian, M., Ranjbar, A. M., Amraee, T., & Shirani, A. R. (2007). Optimal placement of phasor measurement units: Particle swarm optimization approach. *2007 International Conference on Intelligent Systems Applications to Power Systems* (*1*), (pp. 1–6), Nov.

[HSA12] Hassan, M. Y., Suharto, M. N., Abdullah, M. P., Majid, M. S., & Hussin, F. (2012). Application of particle swarm optimization for solving optimal generation plant location problem. *International Journal of Electrical and Electronic Systems Research, 5*.

[HSR08] Horowitz, E., Sahni, S., & Rajasekaran, S. (2008). *Computer algorithms*. Summit, NJ: Silicon Press.

[HYB08] Hetzer, J., Yu, D. C., & Bhattarai, K. (2008). An economic dispatch model incorporating wind power. *IEEE Transactions on Energy Conversion, 23*, 603–611. doi:10.1109/TEC.2007.914171.

[JAW+10] Javed, F., Arshad, N., Wallin, F., Vassileva, I., & Dahlquist, E. (2010). Engineering optimization models at runtime for dynamically adaptive systems. *IEEE International Conference on Engineering of Complex Computer Systems, Singapore* (pp. 253–254).

[JMF99] Jain, A., Murty, M., & Flynn, P. J. (1999). Data clustering: A review. *ACM Computing Surveys, 31*(3), 264–323.

[JW00] Jennings, N., & Wooldridge, M. (2000). Agent-oriented software engineering, *Proceedings of the 9th European Workshop on Modeling Autonomous Agents in a Multi-Agent World, Spain*.

[JTM10] Jupe, S. C. E., Taylor, P. C., & Michiorri, A. (2010). Coordinated output control of multiple distributed generation schemes. *IET Renewable Power Generation, 4*, 283. doi:10.1049/iet-rpg.2009.0142.

[Knu97] Knuth, D. (1997). *The art of computer programming, vol. 1: Fundamental algorithms*. Westford, MA: Pearson Education, Inc.

[Kad09] Kadar, P. (2009). Multi objective optimization of smart grid structure. *Proceedings of 15th International Conference on Intelligent System Applications to Power Systems, Greece*.

[Wie01] Wiegers, K. *Software requirements* (Second edition). City, ST: Microsoft Press, ISBN-10-0-7356-1879-8.

[KH09] Karnouskos, S., & Holanda, N., Simulation of a smart grid city with software agents, *Proceedings of European Modeling Symposium (EMS 2009), Athens, Greece*, November 2009.

[KHS05] Khattam, W., Hegazy, Y., & Salama, Y. (2005). An integrated distributed generation optimization model for distribution system planning. *IEEE Transactions on Power Systems, 20*(2), 1158–1165.

[KJN+04] Kirschen, D., Jawayeera, D., Nedic, D., & Allan, N. (2004). A probabilistic indicator of system stress. *IEEE Transactions on Power Systems, 19*(3), 1650–1657.

[NP] Nygard, K., & Ranganathan, P. A LP based large scale decomposition and optimization in smart grid.

[KP94] Kozina, G., & Perepelista, V. (1994). Interval spanning trees problem: Solvability and computational complexity. *Interval Computations, 1*, 42–50.

[kro63] Kron, G. (1963). *Diakoptics: The piecewise solution of large-scale systems* (Vol. 2).

[Kru56] Kruskal, J. (1956). On the shortest spanning subtree of a graph and the travelling salesman problem. *Proceedings of the American Mathematical Society, 7*(1), 48–50.

[Kru06] Krutz, R. (2006). *Securing SCADA systems*. City, ST: Wiley.

[KY97] Kouvelis, P., & Yu, P. (1997). *Robust discrete optimization and its applications*. City, MA: Kluwer Academic Publishers.

[LAP06] Ling, J. M., Aughenbaugh, J. M., & Paredis, C. J. (2006). Managing the collection of information under uncertainty using information economics. *ASME Journal of Mechanical Design, 128*(4), 980–990.

[LSC05] Litos Strategic Communication. (2005). *The smart grid: An introduction* [Online]. Retrieved from http://energy.gov/oe/downloads/smart-grid-introduction-0.

[LW66] Lawler, L., & Wood, E. (1966). Branch-and-bound methods: A survey. *Operations Research, 14*(4), 699–719.

[Mic15] Microsoft. (2015). *LINQ (Language-integrated query)* [Online]. Retrieved from http://msdn.microsoft.com/en-us/library/vstudio/bb397926.aspx.

[MB03] Member, S., & Begovic, M. (2003). Nondominated sorting genetic algorithm for optimal phasor measurement placement. *IEEE Transactions on Power Systems, 18* (1), 69–75.

[MEG11] Hall, M., Frank, E., Holmes, G., Pfahringer, B., Reutemann, P., & Witten, I. H. (2009). The WEKA datamining software: An update. *SIGKDD Explorations, 11*(1), 10–18.

[MG05] Montemanni, R., & Gambardella, L. (2005). A branch and bound algorithm for the robust spanning tree problem with interval data. *European Journal of Operational Research, 161*(3), 771–779.

[MH12] Mahaei, S. H., & Hagh, M. T. (2012). Minimizing the number of PMUs and their optimal placement in power systems. *Electric Power Systems Research, 83*(1), 66–72.

[Mom01] Momoh, J. (2001). *Electric power system Applications of optimization*. City, NY: Marcel Dekker.

[Moo91] Moore, R. (1991). Global optimization to prescribed accuracy. *Computers and Mathematics with Applications, 21*(6), 25–39.

[MMK10] Aminifar, F., Member, S., Khodaei, A., Fotuhi-firuzabad, M., & Member, S. (2010). Contingency-constrained PMU placement in power networks. *IEEE Transactions on Power Systems, 25*(1), 516–523.

[MMM+13] Mazhari, S. M., Member, S., Monsef, H., & Lesani, H. (2013). A multi-objective PMU placement method considering measurement redundancy and observability value under contingencies. *IEEE Transactions on Power Systems, 28*(3), 2136–2146.

[MWJ+10] Mohsenian-Rad, M., Wong, V., Jatskevich, J., & Schober, R Optimal and autonomous incentive-based energy consumption scheduling algorithm for Smart grid. *Proceedings in IEEE Power Engineering Society Conference on Innovative Smart grid Technologies*, January 2010 (pp. 1–8).

[NCK+09] Nasri, A., Conejo, A. J., Kazempour, S. J., & Ghandhari, M. (2014). Minimizing wind power spillage using an OPF with FACTS devices. *IEEE Transactions on Power Systems*, 1–10. doi:10.1109/TPWRS.2014.2299533.

[New06] Newman, M. E. J. (2006). Modularity and community structure in networks. *Proceedings of the National Academy of Sciences of the United States of America, 103* (23), 8577–8582.

[New04] Newman, M. E. J. (2004). Fast algorithm for detecting community structure in networks. *Physical Review. E, Statistical, Nonlinear, and Soft Matter Physics, 69* (6), 1–5.

[NF12] Nguyen, C., & Flueck, A. (2012). Agent based restoration with distributed energy storage support in smart grids. *IEEE Transactions on Smart Grid, 3*(2), 1029–1038.

[NGL+11] Nygard, K., Ghosn, S., Loegering, D., McCulloch, R., & Ranganathan, P. (2011). Implementing a flexible simulation of a self-healing smart grid. *2011 International Conference on Modeling, Simulation and Visualization Methods, Nevada.*

[NPF05] Nuqui, R. F., Phadke, A. G., & Fellow, L. (2005). Phasor measurement unit placement techniques for complete and incomplete observability. *IEEE Transactions on Power Delivery, 20*(4), 2381–2388.

[NRG+11] Nygard, K., Ranganathan, P., Ghosn, S., Loegering, D., McCulloh, R., & Chowdhury, M. Optimization models for energy reallocation in a smart grid. *IEEE Machine to Machine Communications and Networking Workshop, Shanghai, China, April 2011* (pp. 186–191).

[NS02] Nagata, T., & Sasaki, H. (2002). A multi-agent approach to power system restoration. *IEEE Transactions on Power Systems, 17*(2), 457–462.

[NYI] Retrieved from http://www.nyiso.com/public/markets_operations/market_data/load_data/index.jsp.

[NW88] Nemhauser, G., & Wolsey, L. (1988). *Integer and combinatorial optimization*. City, NJ: Wiley.

[PFR09] Pipattanasomporn, M., Feroze, H., & Rahman, S. Multi-agent systems in a distributed smart grid: Design and implementation. *Proceedings of IEEE PES 2009 Power Systems Conference and Exposition, Seattle, Washington, USA, March 2009* (pp. 1–6).

[RAK07] Raghavan, U. N., Albert, R., & Kumara, S. (2007). Near linear time algorithm to detect community structures in large-scale networks. *Physical Review. E, Statistical, Nonlinear, and Soft Matter Physics, 76*(3), 1–12.

[RN12] Ranganathan, P., & Nygard, K. A smart agent oriented linear programming control in electric grid. *Annual Electric Power and Energy Conference, Canada, October 2012* (pp. 102–106).

[Ruo13] Ruohonen, K. (2013). *Graph theory* (p. 108).

[PR] *Report from Pike research*. Retrieved from http://www.pikeresearch.com/research/smart-grid-data-analytics.

[PPM+08] Padgham, Parkes, Müller & Parsons (Eds.) The permutable POMDP: Fast solutions to POMDPs for preference elicitation, F. Doshi, N. Roy, *Proceedings of 7th International Conference on Autonomous Agents and Multiagent Systems (AAMAS 2008), May, 12–16, 2008, Estoril, Portugal* (pp. 493–500).

[PSN+88] Powell, W., Sheffi, Y., Nickerson, S., Butterbaugh, K., & Atherton, S. (1988). Maximizing profits for North American Van Lines truckload division: A new framework for pricing and operations. *Interfaces, 18*(1), 21–41.

[Pow98] PowerGen plc. (1998). *Private communication.*

[RI10] Rahimi, R., & Ipakchi, A. 2010. Demand response as a market resource under the smart grid paradigm. *IEEE Transactions on Smart Grid, 1*(1), 52–66.

[Qui92] Quinlan, J. (1992). Learning with continuous classes. In A. Adams & L. Sterling (Eds.), *Proceedings of the Artificial Intelligence* (pp. 343–348).

[QH08] Qiao, W., & Harley, R. G. (2008). Grid connection requirements and solutions for DFIG wind turbines. In *IEEE Energy 2030 Conference 2008. IEEE* (pp. 1–8). doi:10.1109/ENERGY.2008.4781068.

[RN10] Ranganathan, P., & Nygard, K. (2010). An optimal resource assignment problem in Smart grid. *The Second International Conference on Future Computational Technologies and Applications, Portugal, November 26* (pp. 75–82).

[RN12] Ranganathan, P., & Nygard, K. (2012). A smart agent oriented linear programming (AOLP) architecture in electric grid. *2012 Annual Electric Power and Energy Conference, London, Canada, October 10–12.*

[RPT07] Rahman, S., Pipattanasomporn, M., & Teklu, Y. (2007). Intelligent distributed autonomous power systems (IDAPS). *Proceedings of IEEE PES Annual General Meeting, Tampa, Florida, USA.*

[Som11] Sommerville, I. (2011). *Software engineering*. Boston, MA: Pearson Education, Inc.

[Sal04] Salam, S. (2004). Comparison of Lagrangian relaxation and truncated dynamic programming methods for solving hydrothermal coordination problems. *Proceedings of International Conference on Intelligent Sensing and Information Processing* (pp. 265–270).

[SAS] SAS. *About SAS* [Online]. Retrieved from http://www.sas.com/company/about/index.html.

[SAS10] SAS. (2010). *SAS/OR(R) 9.2 user's guide: Mathematical programming* [Online]. Retrieved from http://support.sas.com/documentation/cdl/en/ormpug/59679/HTML/default/viewer. htm#netflow_sect63.htm.

[SAP] SAS. *PROC NETFLOW statement* [Online]. Retrieved from http://support.sas.com/documentation/cdl/en/ormpug/63352/HTML/default/ormpug_netflow_sect016.htm.

[SCJ13] Sudhakar AVV, Chandram, K., Jayalaxmi, A. (2013). Multi area economic dispatch with tie line loss using secant method and tie line matrix. *International Journal of Applied Power Engineering (IJAPE), 2*(3), 115–124.

[SBA13] Sahu, B., Lall, A., Das, S., & Manoj Kumar Patra, T. (2013). Economic load dispatch in power system using genetic algorithm. *International Journal of Computer Applications, 67*(7).

[SFN14] Sánchez-García, R., Fennelly, M., Norris, S., Wright, N., Niblo, G., Brodzki, J., & Bialek, J. (2014). Hierarchical spectral clustering of power grids. *IEEE Transactions on Power Apparatus and Systems, 29*(5), 2229–2237.

[SFO13] Shiokawa, H., Fujiwara, Y., & Onizuka, M. (2013). Fast algorithm for modularity-based graph clustering. *Proceeding Twenty-Seventh Conf. Artif. Intell.* (pp. 1170–1176).

[Sol02] Solomatine, D. P. (2002). Data-driven modelling: paradigm, methods, experiences. *Proceedings of 5th International Conference on Hydroinformatics, Cardiff, UK.*

[SD03] Solomatine, D. P., & Dulal, K. N. (2003). Model tree as an alternative to neural network in rainfall-runoff modeling. *Hydrological Science Journal, 48*(3), 399–411.

[She95] Shenker, S. (1995). Fundamental design issues for the future internet. *IEEE Journal of Select Areas Communication, 13*(7), 1176–1188.

[Son99] Song, Y. (1999). *Modern optimization techniques in power systems*. City, ST: Kluwer Academic Publishers.

[SMD0+9] Smith, J. C., Milligan, M. R., DeMeo, E. A., & Parsons, B. (2007). Utility wind integration and operating impact state of the art. *IEEE Transactions on Power Apparatus and Systems, 22*, 900–908. doi:10.1109/TPWRS.2007.901598.

[SatO6] Sathyajith, M. (2006). *Wind energy: Fundamentals, resource analysis and economics*. City, ST: Springer.

[SP10] Russell, S., & Norvig, P. (2010). *Artificial intelligence: A modern approach*. Upper Saddle River, NJ: Pearson Education, Inc.

[SS85] Sullivan, R., & Secrest, C. (1985). A simple optimization DSS for production planning at Dairymans Cooperative Creamery Association. *Interfaces, 15*(5), 46–53.

[SSP12] Saha Roy, B. K., Sinha, A. K., & Pradhan, A. K. (2012). An optimal PMU placement technique for power system observability. *International Journal of Electrical Power & Energy Systems, 42(1)*, 71–77.

[SV10] Saber, A. Y., & Venayagamoorthy, G. K. (2010). Intelligent unit commitment with vehicle-to-grid – A cost-emission optimization. *Journal of Power Sources, 195*(3), 898–911.

[SVY10] Saber, A. Y., & Venayagamoorthy, G. K. (2010). Efficient utilization of renewable energy sources by gridable vehicles in cyber-physical energy systems. *IEEE Systems Journal, 4*(3), 285–294. doi:10.1109/JSYST.2010.2059212.

[Tem91] Templeman, B. (1991). *Optimization and decision support systems in civil engineering*. City, ST: Routledge, Gordon and Breach Science Publishers Ltd.

[UWD16] University of Washington Department of Electrical Engineering. (2016). *Power systems test case archive* [Online]. Retrieved from http://www.ee.washington.edu/research/pstca/.

[XCF09] Xie, L., Carvalho, P. M. S., Ferreira, L. A. F. M., Krogh, B. H., Popli, N., & Ilić, M. D. (2011). Wind integration in power systems: Operational challenges and possible solutions. *Proceedings of the IEEE, 99*, 214–232. doi:10.1109/JPROC.2010.2070051.

[Wad83] Waddell, R. (1983). A model for equipment replacement decision and policies. *Interfaces, 13*(4), 1–7.

[Wan01] Wang, H. (2001). Multi-agent co-ordination for the secondary voltage control in power system contingencies. *Proceedings of IEEE Generation, Transmission and Distribution, 148*(1), 61–66.

[Web01] Mathworks, Inc. *Binary Integer Programming* (BINTPROG). Retrieved August 31, 2010, from http://www.mathworks.com.

[Web02] Retrieved August 31, 2010, from http://www.oe.energy.gov/smartgrid.htm.

[WE] Retrieved from http://www.cs.waikato.ac.nz/ml/weka/.

[Web97] Weber, J. D. (1997). *Implementation of a Newton based optimal power flow into a power system simulation environment*. University of Illinois at Urbana-Champaign.

[WG10] Wang, C., and Groot, M. Managing end-user preferences in the Smart grid. *ACM International Conference on Energy efficient computing and networking, Germany, April 2010* (pp. 357–363).

[Wil93] Williams, P. (1993). *Model building in mathematical programming (3rd revised edition)*. Chichester, England: Wiley.

[WL13] Wen, M. H. F., & Li, V. O. K. (2013). Optimal PMU placement for wide-area monitoring using chemical reaction optimization. *2013 I.E. PES Innovative Smart Grid Technologies Conference* (pp. 1–6), Feb.

[Wol98] Wolsey, L. (1998). *Integer programming*. City, NY: Wiley.

[Wri97] Wright, S. (1997). *Primal-dual interior point methods*. SIAM, 145–157.

[ZRK07] Zarei, M., Roozegar, A., Kazemzadeh, R., & Kauffmann, J. M. (2007). Two area power systems economic dispatch problem solving considering transmission capacity constraints. *Power, 12*, 12.

Index

P. Ranganathan, K.E. Nygard, *Distributed Linear Programming Models in a Smart Grid*, Power Electronics and Power Systems, DOI 10.1007/978-3-319-52617-1

The manufacturer's authorised representative in the EU is Springer Nature Customer Service Centre GmbH, Europaplatz 3, 69115 Heidelberg, Germany. If you have any concerns regarding our products, please contact ProductSafety@springernature.com

Printed and bound by CPI Group (UK) Ltd, Croydon, CR0 4YY

30/09/2026

02225221-0002